Biblical Homestead Community and Traditions
PrairieDustTrail.com Magazine
Volume 1 - 2023 Edition

ISBN: 9798866321278
Independently published

Cover by Danielle Tate

Authors:

Dawnita Fogleman,
Publisher & Editor-in-Chief

Myrna J Buckles, *Senior Editor*
Kathryn (Fogleman) White, *Copy Editor*
Christayla Vassar, *Associate Editor*

Contributors:

Damian Allen	Patricia Jean Pitonyak	Christina Thompson
Kellie Doiron	Francis Roland	Harold Thornbro
Prairie Ruth Fogleman	Tresa Salters	Ryland Vassar
Deborah Hanyon	Elizabeth Santelman	Julie Voth
Mike Hein	Kristen Smith	Hope Ware
Andrea Hutchison	Snakes on the Prairie	Israel Wayne
Emmie Manor	Danielle Tate	Steven White
Ken McKinley	Wyatt Tate	

Photos:

- Elizabeth Santelman - SunshineinMyNest.com (Summer 2023 cover)
- Abagail Gray - FaithHasMade.com (Fall 2023 cover)
- Diana Bristol - Dbristolmk@cox.net
- Misc. photos & original clipart from the Fogleman Farm
- Copyright Free - Faithful photographic reproductions of original two-dimensional works of art from the public domain in their countries of origin and areas where the copyright term is the author's life plus 70 years or fewer.

Are you a writer, photographer, or aspiring creative?

If you would like to contribute to the Prairie Dust Trail magazine, here's what you need to know:

- **Core Values:** Faith, Consistency, Resilience, Community
- **Goals:** To provide value, inspiration, and entertainment. Connect readers with pioneer skills from the past. Help them become more prepared for the inevitable dust storms of life. Promote more sustainable living, learning to set good boundaries on time, energy, and resources through routines, habits, and proper tools.

For the full set of guidelines, please email Dawnita@PrairieDustTrail.com.

Disclosures and disclaimers:

The contents of this magazine are for information and entertainment purposes only. Always check with your healthcare provider for any health concerns before implementing any advice or ideas presented. The FDA has not evaluated the statements in this publication. These articles are not intended to diagnose, treat, cure, or prevent any disease.

Unless otherwise noted, none of the information in this publication should take the place of professional legal advice. You should always do your own research and consult your own national, state, and local authorities, ordinances, or laws.

Unless otherwise noted, none of the information in this publication should take the place of your veterinarian's advice. Always consult your veterinarian for the care of your animals.

Purchasing products through an affiliate link contributes to the success of this magazine and its contributors at no extra cost to you. We only recommend products or services we use personally or believe are valuable to readers. We are disclosing this in accordance with the FTC's 16 CFR, Part 255: "Guides Concerning the Use of Endorsements and Testimonials in Advertising."

Table of Contents

Echo from the Past

Editor's Epistle	8
Goal Setting for the New Year	10
Salt Gathering of the Past	12
Bless Your Children! (And a Blessing!)	15
Back To The Roots Of Frugal Homesteading	18
Pioneer Weekly Chore List	22
Once upon a time, prepping was NOT an option	24
Preparing Your Garden Beds for Spring	26
The Importance of Beef Labeling	28
Need Seeds? Where to Find Them	31
Just the Facts - Politics from the Horses' mouths...	34
My first beaver trapping experience	35
Homestead Act	37
Recipe Corner	39
Weather Modification and Geoengineering	40
How Jesus stretched me to see the truth about yoga	42
101 Ways to Kill Sourdough	46

Fostering Community

Editor's Epistle   48

Forging kinship through struggle and
differences   49

Starting Seeds   51

Poultry Care   54

Quail Basics and some more   55

Building Communities   58

5 Plants to Forage in Spring   61

Practical Rabbit Care Tips and Tricks   64

Engagement Nurtures Lasting Community   65

The Importance of Teaching Community to
Your Children   67

Death by a Thousand Cuts:
The Incremental Path towards Food
Destruction   70

5 Reasons Blueberries are Superfoods   73

What does it mean to know thyself?   74

Recipes   76

Just the Facts - Politics from the horse's mouth   77

Fostering Community -
Three Case Studies & What We Can Learn From
Them   79

The Adventure of Homeschooling

Editor's Epistle	86
The Manual for Children	87
Snake Identification and Care for Bites	89
Homeschooling in a Shoe	91
What is Turmeric Good For?	96
Financing Homeschool Frugally	98
Essential Tools for the Homestead	101
Recipes	103
4 Reasons Every Homestead Should Have Quail	104
Homeschooling 101	106
The Importance of Mentorship	108
What I Wish I Would Have Done Differently as a Homeschooled Teen	110
If I Were A Pioneer Mother, Here's How I'd Guide You	112
Lone Wolf Prepper	116
You Can Grow Food... ANYWHERE!	119
Just the Facts-Politics from the horse's mouth...	123

Passing on Traditions

Editor's Epistle 125

Breaking the Curse of Generational Trauma: A Path to Healing 126

Christian Herbalism 130

My Window into the Past - Passing on Traditions 133

Sourcing Homestead Items Cheaply 136

Passing on Traditions 138

A Republic of Good Behavior vs. A Democracy of Tyrannical Nobles 141

3 Superfoods that are Truly Healthy 143

Recipes 145

5 Easy Steps to Transition to a Whole Food Plant-Based Way of Eating 147

Food for the Apocalypse 149

Just the Facts - Politics from the horse's mouth... 153

Foreign Land Ownership Concerns Spark Changes on Congressional Committee 154

How to Stay Connected During a Crisis 155

The Best Gifts for New Moms 157

Preparing for a VBAC 161

Gift Ideas for Preppers 165

4-H History Project Ideas 166

From Struggle to Strength: High School and Learning Disabilities 168

Echo from the Past

Winter 2023 PrairieDustTrail.com Volume 1 Issue 1

PrairieDustTrail.com

Connect with the past ~ Prepare for the future ~ Live more sustainably

Plan for the year
Prep for spring
Pass on skills
Praise through trials

An Echo from the Past

Editor's Epistle

It's 2023 and boy howdy have the past several years been something else. With all the craziness though, I've found a lot to be thankful for. People have become more aware and started questioning things, rather than taking everything they are told at face value.

In my heart, I praise the Lord for this change in mindset. I hate that so much hardship, hatred, and loss have driven people to this place. I also know sometimes, that's what it takes. I'd rather not count the times our Father has had to figuratively drag me, kicking and screaming toward some blessing I didn't understand. The times HE has opened my eyes to something I didn't really want to see, or my ears to something I definitely didn't want to hear.

> *"And we know that all things work together for good to them that love God, to them who are called according to his purpose." Romans 8:28*

This verse doesn't mean HE won't give us more than we can handle. As a matter of fact, HE often overloads us so that we will continue to lean on Him. Rather, HE promises we will not be tempted beyond what we can handle and that HE'll provide a getaway.

> *"There hath no temptation taken you but such as is common to man: but God is faithful, who will not suffer you to be tempted above that ye are able; but will with the temptation also make a way to escape, that ye may be able to bear it." 1 Corinthians 10:13*

Spiritual growing pains sometimes hurt every bit as bad as a child with leg cramps. It's hard sometimes to leave the blissful ignorance behind and embrace the Truth of Scripture.

> *"For the word of God is quick, and powerful, and sharper than any two-edged sword, piercing even to the dividing asunder of soul and spirit, and of the joints and marrow, and is a discerner of the thoughts and intents of the heart." Hebrews 4:12*

The more willing we are, the easier the process.

Like childbirth, when a woman fights the contractions, they definitely hurt more than when she breathes through them and allows her body to work as it's meant to.

On this new journey of publishing this magazine, I'm addressing a few issues I've seen rise up over these past several years.

1. While there are many new and wonderful homesteading teachers now, so many of them are new themselves. I applaud them for taking the leap into becoming more self-sufficient and sharing their journey. Yet, some of those following them find things to be very different once they take the leap themselves. Homesteading can be overwhelming and frustrating, as well as rewarding.

2. Few homesteading resources are steeped in Scripture. Let's face it, we can't embark on this adventure without the hand of God in every aspect of our lives. HE is the one who ultimately sustains us.

3. Encouragement and accountability are so important. Our natural tendencies lean toward either completely overthinking or jumping in half-cocked. (Or a mixture of completely discombobulated and unbalanced mess!)

4. Homesteading isn't a set-in-stone, one-size-fits-all, textbook approach. It's an adventure lifestyle that moves in waves and seasons directed by weather patterns, governmental influence, and personal journeys.

As we move through the issues of this year in the editions of this magazine, I ask you, Dear Reader, to pray for the contributors and advertisers. This is bound to be a year of growth and learning for us all.

In the meantime, I pray you are each blessed beyond measure and find many helpful and inspiring ideas to encourage you on your journey.

Many Blessings,

Dawnita

Dawnita Fogleman is a fifth-generation Oklahoma Panhandle homesteader, homeschool mom of six, Meemaw, author, artist, and award-winning journalist. At PrairieDustTrail.com she connects today's pioneers with the past to live more sustainably, weathering the dust storms of life with God's Divine help.

Goal Setting for the New Year

By Christayla Vassar

As people, we like to see progress toward the things we want out of life. We feel annoyed and stuck when our life doesn't go in the direction we have planned. Though there's no true way to know what twists and turns our path will take, we can rely on God. He said:

> *"For I know the thoughts that I think toward you, saith the Lord, thoughts of peace, and not of evil, to give you an expected end. - Jeremiah 29:11"*

When we feel God leads us in a certain direction, we can make goals for that. For example, maybe God asks you to lead a Bible study. It's your job to prepare the place and the time, and to research the topic He's put on your heart. You can invite people too, but He will take care of who takes part.

Dr. Benjamin Hardy says we need to commit to up to three areas in our life to grow in for a season. This could be a real season, a year, or even several years. Otherwise, we stretch ourselves too thin.

First, let's begin setting our goals by figuring out where God has been asking us to grow. What are those areas of your life He's led you to?

These might be: Finances, Family/Home, Learning/Education, Writing, Business, Health, Work, Faith...

These will be the three priorities for you to set goals under this year. Let's take three examples: Family/home, Business, and Homestead.

To achieve a closer or more intentional family, you need to make a priority and goal of it. Break down your priority into smaller goals. For an intentional family, maybe you have pizza and games every Friday night. Then you take Sunday as a day to go to church and then volunteer together.

Let's say you meal plan to keep your home running smoothly and your family fed. To do that you plan, shop, and even meal prep one day a week.

For the priority of business, the goals may seem much bigger and harder to achieve. You want to gain 150,000 followers or gain 30 more clients or customers. This could seem like an insurmountable challenge.

These kinds of goals can seem like elephants. So, how do you eat an elephant? One bite at a time. Make your gigantic goals into smaller ones. Schedule them. Do the work. Then achieve them.

This means writing out and scheduling those social media updates, emails, and blog posts. It means getting out there and talking with people to help them. It means offering your offer instead of being too scared of a no.

For the priority of the homestead, things will happen to set you back. You may be working on a chicken coop and the weather doesn't cooperate so you can get it done in a timely manner. Your chickens and ducks may get swiped by bobcats or hawks. They might die of Marek's disease, bumblefoot, or egg binding. Your goats and pigs might escape and eat your vegetable garden. Plus, your cow may decide the grass is greener in other pastures.

Don't let this keep you from hitting your goals. Your goal may be to grow enough tomatoes for 25 quarts of stewed tomatoes. Do it. You may decide to make all your dairy products at home. Learn them, make them, eat them, and enjoy them.

Setbacks hit hard. Please remember this: don't only acknowledge when you hit your goals. Stop to celebrate the progress you're making toward your goal. Some people don't even stop to celebrate achieving their goals. They wish they were still further along. Don't be one of them.

Remember to revisit your goals every day and revisit your progress every week. If a priority must change or a goal isn't achievable because of life circumstances, don't sweat it. Rev-evaluate and change it if you need to.

You can do this. Write those goals and get to making them happen with God's guidance and blessing.

Christayla Vassar is a wife, homeschool mom, and entrepreneur. She's been gardening ever since she can remember and received her Permaculture Design Certificate in June 2022. She enjoys helping parents forge their own unique family design with The Folksy Side of Life. https://www.thefolksysideoflife.com/

Salt Gathering of the Past

By Myrna Buckles

Salt is a naturally occurring mineral, essential for human and animal life. It is a commodity in the world of trade as well as a seasoning and preservative for foods.

Historically, salt has been evaporated from seawater and from salt mines. In areas where salt has been difficult to obtain (not near the ocean or salt plains), salt has been evaporated from rivers, streams, and springs.

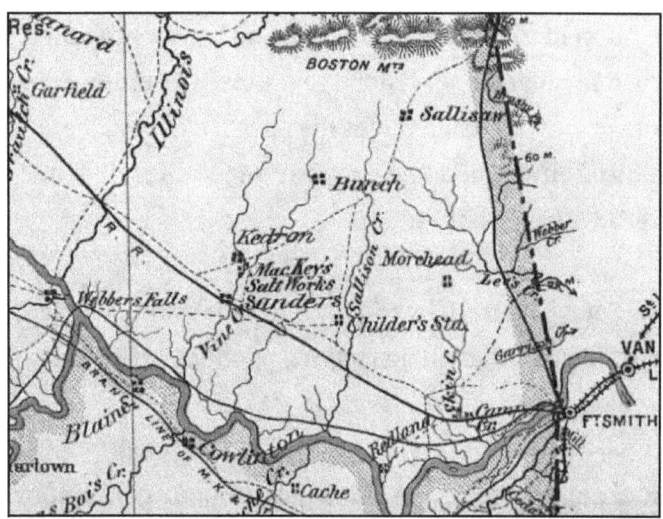

In the early 1800s, there was no going to the corner grocery store to pick it up or ordering it online. In Eastern Oklahoma (before it was a territory or state), there were two known sites developed on the Illinois River, one operated by the Beans and later Walter Webber who was Cherokee, and one operated by Samuel Mackey. Salt evaporation was a principal industry in the area at that time.

Samuel Mackey, whose family had been removed with the Cherokee from Alabama to Arkansas in 1819 and from Arkansas to Oklahoma in 1828, developed a considerable establishment to evaporate salt from a "spring that discharged salt water through an opening in the solid rock in the bottom of the Illinois River".(1) They installed a pipe to carry the spring water to the surface of the river and a pump to carry the water to the kettles on the banks of the river. The kettles weighed about 1500 pounds each.

There was a furnace under the kettles that was used to boil the water, evaporate it, and eventually harvest the salt. Can you imagine how big those kettles were? They had to be quite large to weigh 1500 pounds.

Samuel Mackey shipped salt up and down the Illinois River supplying a large demand and furnished lodging and food to travelers who passed that way. "He died in 1839 and his sons, James and W.T. succeeded to his business." (1)

These sites were turned over to Cherokee Tribal members (Webber and Mackey) following their removal from Arkansas to Oklahoma Territory.

While Mackey and others were making salt for the masses, women were surviving as best they could with what they had. One such story recorded in the Chronicles of Oklahoma is below:

> *"While our men ancestors were gone somewhere to get salt, our women dug up the smokehouse dirt floors where salty meat had hung and dripped. They boiled this in huge iron wash-pots and strained the water and evaporated it to get salt." (2)*

This is a description of a similar facility in Kentucky around the same time frame and gives us an idea of how big of an operation these sites could be:

At this time they had one furnace with 40 kettles, which six to 10 men operated. Forty kettles produced from 25 to 30 bushels of salt per day. This was shipped to settlements along the Missouri River and to St. Louis, where it sold for $2 to $2.50 per bushel.

The salt works were later expanded by enlarging the existing furnace and erecting a new one, enabling each furnace to hold sixty kettles. Each furnace produced about 100 bushels of salt per day. According to Boone, 300 gallons of water were required to make one bushel of salt. Keeping so many kettles going required twice as many men. (3)

Myrna Stiles Buckles grew up in Southeastern Oklahoma at the Gardner Mansion & Museum. History was a family passion with both of her parents serving on the Oklahoma Historical Society Board of Directors. She gave tours of the mansion & museum throughout her middle school, high school & college years. In later years, she recorded family stories and helped her father research subjects like the salt works, Indian Fishing methods, and the eradication of brucellosis in Oklahoma. Myrna is a retired Public Health Service Officer who served Native American Tribes in Arizona, California, North & South Dakota, Iowa & Nebraska. She has a heart for service and has led many short-term mission teams to Nicaragua. She has been called by God to empower women in impoverished communities in Nicaragua. She & her husband recently moved to Nicaragua to fulfill that calling. You can find more at MyrnaStilesBuckles.com.

1. Oklahoma Historical Society. Chronicles of Oklahoma, Volume 10, Number 4, December 1932, periodical, December 1932; Oklahoma City, Oklahoma. https://gateway.okhistory.org/ark:/67531/metadc1827194/: accessed January 8, 2023),

2. The Gateway to Oklahoma History, https://gateway.okhistory.org; crediting Oklahoma Historical Society.

3. Oklahoma Historical Society. Chronicles of Oklahoma, Volume 35, Number 4, 1957, periodical; Oklahoma City, Oklahoma. https://dc.library.okstate.edu/digital/collection/p17279coll4/id/39024/

4. Boones laid claim to Mackay's Lick by WARREN DALTON AND DEBORAH THOMPSON, The Columbia Tribune, Published July 6, 2014, https://www.columbiatribune.com/story/lifestyle/around-town/2014/07/06/boones-laid-claim-to-mackay/21734443007/

Bless Your Children! (And a Blessing)

By Christayla Vassar

Do you ever bless your children? Maybe you've thought about it, but it seems like something they only did in Biblical times.

You can give your children two types of blessings. One type happens in your prayers. The other happens actively, with you speaking blessings over them. This article focuses on the second type.

You can do this at any time of the day, but I love to bless them in the morning. There's just something about connecting with them in the morning and giving them a blessing to go with them as they go about their day. But not everyone has time in the morning. You may find it works better for your family at bedtime instead.

You can put your hand on your child's head, hold their hand, or hold them in your lap. Whatever works best to bless them.

What do you bless your children with? You can ask God to bless your child with:

Grace

Pray for God's gift of grace for your children. We're all sinners and make mistakes. Pray your children will accept His gift of Grace as they grow. (Romans 3:23-24)

Wisdom

Our children need wisdom. They make decisions on obeying. They decide on interactions with siblings and friends, sharing toys, and more. We want them to make decisions that honor God. Especially since as they grow, they will need to make more difficult decisions. Pray over their decisions, and ask God to bless them with wisdom to make them. (Proverbs 2:6)

Straight Paths

Your child takes many steps every day. Bless their steps. Ask God to keep them on the straight and narrow path. Bless their steps forward each day as they walk with God. (Proverbs 3:5-6)

Heart

This world embraces the ungodly. Bless your child's heart so that they may turn away from the temptations and see the broken and the hurting. (Luke 6:45)

Peace

This world can hurt, especially as we grow older and experience the evil that's within it. Bless them with the peace that surpasses understanding, God's peace. (Philippians 4:7)

The Fruits of the Spirit

What are the fruits of the Spirit? Love, joy, peace, patience, kindness, goodness, faithfulness, gentleness, and self-control. (Galatians 5:22-23) Our children need each of these. Pray for God to bestow these on your children.

Protection

God protects us from evil and Satan and his demons. Bless your children with this protection, asking God to cover them. (Deuteronomy 31:6)

A Thirst for HIM

We want our children to pursue God and hunger and thirst for His Word. Ask God to bless them with it. (John 4:14)

You can also ask God to give your children:

Prosperity

Prosperity means more than enough. God will bless us with enough if we will receive it, but he wants us to prosper. He wants us to invest what He provides to us and bless others with it generously. Bless that their hearts will receive God's provision and they will share it generously. (Malachi 3:10)

Health

So many people choose to eat what tastes good versus what is healthy. In addition, so many able-bodied people choose to sit versus get the exercise they need. Bless your children with the wisdom to make choices on what will nourish them and their bodies so they can further God's kingdom. (1 Corinthians 1:31)

Here is a blessing you can use. Feel free to make it your own by changing it up. Or write your own!

"May God bless you this day with His grace and goodness. May He guide your feet on the straight path and raise your eyes to heavenly things. May He help you guard your heart against temptation. Embrace the wisdom he bestows upon you, dearest child. May you cultivate the Fruits of the Spirit in your everyday life. No matter what the world throws at you, may you rest in His peace always. Watch over Your child, Lord, as he/she grows, and lead him/her on Your path of righteousness. Keep him/her under Your hedge of protection and may all evil done against them turn to goodness and glorify you. Comfort him/her when this world fails him/her by reminding him/her that heaven awaits. And when he/she fails You, remind him/her of Your never-ending forgiveness. May you, dear child, thirst for God and His Word. May you ever seek to know Him and follow His ways. May God bless you with prosperity, generosity, and health as you follow His commandments. AMEN."

Back To The Roots Of Frugal Homesteading

By Harold Thornbro

"There is treasure to be desired and oil in the dwelling of the wise; but a foolish man spendeth it up." Proverbs 21:20

Our God, who provides all things and has provided for us beyond measure, has also instructed us to be mindful of how we use these gifts and not to waste them foolishly.

We live in a society that tends to waste and purchase rather than reuse and repurpose, but frugal homesteading is a God-honoring act as well as a wise decision with many benefits that have seemingly been forgotten by many homesteaders.

So how can we homestead frugally, and what are some potentially free or cheap resources?

1. Soil Amendments

Purchasing soil amendments could cost you hundreds, if not thousands, of dollars over the years. The good news is if you know what to look for, this can be one of the easiest and most available free or cheap resources to get your hands on.

Manure

Finding manure, depending on where you live, is very easy to get your hands on, usually free. Horse ranches are a common source of this valuable resource, they usually have mounds of it that they are trying to get rid of. However, Horse manure isn't the only kind you can look for. Cow, sheep, chicken, and rabbit manure can also be found in abundance in many places and can be good for soil building.

Compost

Many counties have free or cheap compost via yard waste centers where folks bring their leaves, grass clippings, and other organic waste material where it gets put in huge piles, stirred up, and turned into rich compost. These centers usually give the compost away to anyone who wants to come and load it themselves.

However, making your own compost from material on your own homestead is always the best option, as you can be absolutely sure what is in it. Having a compost bin and a kitchen scrap bucket can get you started down this path.

Leaves and grass clippings

Especially in the Fall, getting your hands on bags of free leaves can be extremely easy. Leaves can make a great garden cover mulch for the winter and can be used as leaf mold as it decomposes and also be turned into finished compost over time. Grass clippings also make great organic material for composting but can also be used as animal fodder for your livestock. Just make sure you know who you're getting it from and that it hasn't been sprayed with anything.

Wood Chips

These make great mulch for your garden and around trees and shrubs and can be expensive if you are purchasing them by the bag. This is also possibly available for free or cheap from county highway departments and tree trimming companies. County Highway Departments usually require you to come and get it yourself, but many tree trimming companies will even drop it off at your homestead if they are working in your area.

Top Soil

If you are gardening using raised beds especially, you will discover you need to get your hands on a lot of soil to fill them up, and this can get really expensive really quickly. Many County Highway Departments also have topsoil available for pickup, but another source is excavating companies. Many times when they remove soil from a job site, they will bring it back to their property and pile it up. If soil isn't available for free from these places in your area, just remember that buying it by the cubic yard in bulk is much cheaper than by the bag usually, and it's also usually a more trustworthy source of good soil.

2. Building Materials

Most homesteaders find they become builders of many things on their property. Whether it be a potting bench, raised garden beds, or a chicken coop, most will do these projects themselves or get a family member or friend to help. Doing this work yourself can save you a lot of money but you can save even more by being creative when it comes to getting your hands on some building materials.

Pallets

These are usually available free or very cheap in many places and, depending on what you are building, can be a great source of material.

Check out the raised beds I built using free pallets.

Breaking down pallets into usable lumber can be a lot of work, especially if you don't have the proper tools for the job. Some pallets are treated with dangerous chemicals and used to transport toxic products, so know where they come from and check the codes on the pallets to find out their treated condition.

Used Lumber

Though usually not free unless you can help someone dismantle something in trade for the lumber, used lumber can be purchased at a huge discount. I've personally saved hundreds of dollars over the years by using this resource on my homestead

Blocks, Bricks, and Rocks

These can be another great resource for your homestead for building raised beds, retaining walls, and walkways, and can sometimes be found for free or cheap.

3. Trash

Making use of what most people throw away can be a huge money saver and provide you with many resources around your homestead. You've probably heard the saying that one man's trash is another man's treasure; that's because the one who counts it as treasure can find a way to use it. I have a friend who works as a janitor, and all year he saves the empty toilet paper tubes he gets at work, and in the spring, he uses them to plant seedlings in. Cutting the bottoms out of milk jugs and other plastic bottles to use as mini-greenhouses for individual small plants is another common practice. The list can go on forever, but you can probably imagine the possibilities of using items most people throw away for a special purpose on your homestead.

4. Used Items

I couldn't possibly mention all the possible things you could get your hands on that could be repurposed and used for something on your homestead, but some of the more common things are barrels that can be used for water collection or feed storage or perhaps an aquaponics system, who knows. Old windows can be used to build cold frames or whole greenhouses. Five-gallon buckets are also a great resource around the homestead that can be used for a great many things. However, nearly anything can be used and repurposed for something if you're creative enough.

5. Making From Scratch

Making things from scratch is not only a healthier practice but can be a huge money saver as well. Many of the things you purchase from the store, such as prepackaged food items and household cleaners, can be made yourself for just pennies on the dollar.

These are just a few ideas for finding cheap and free resources for your homestead to help you develop the life you want to live in a God-honoring and responsible way. Happy homesteading, and may our God, who owns all things, bless you abundantly.

Harold Thornbro is a homesteading and permaculture advocate, host of The Modern Homesteading Podcast, and blogs at RedemptionPermaculture.com

Pioneer Weekly Chore List

By Christayla Vassar

Have you heard of the following? These were the chores for each day of the week for our pioneer ancestors. This kept them on track and kept their household in order.

Wash on Monday

The pioneer women would complete washing on Monday. They would wash the clothes on Mondays. This would allow the time for drying, ironing, and mending before the next Sunday came around. They'd use homemade soap made of lye and lard, washing soda to soften the water, and bluing for white clothing.

Iron on Tuesday

Tuesday: ironing day. Dried clothing items would be sprinkled with water and ironed with irons heated on the stove. Few pioneer women would use starch because of the expense.

Mend on Wednesday

The pioneers completed mending on Wednesday. Issac Singer patented the first commercial and successful sewing machine in 1851. Yet, only the wealthy could afford them. Many pioneer women would complete their sewing by hand.

Churn (or Market) Thursday

On Thursday, a pioneer woman would use her time for churning or marketing (or perhaps a combination of). If they lived far away from town, the pioneer woman would take Thursday to churn butter and make cheese. If they lived in town, the pioneer woman would use the day for going to the general store or market.

Clean on Friday

Pioneer women would clean on Fridays. Though she cleaned her home throughout the week, this day brought deep cleaning. She'd clean the mattresses, the fireplace, the windows, and more, depending on the time of the year.

Bake on Saturday

Saturday would bring delightful smells from the pioneer home. The pioneer women would make bread for the week. They'd prepare Sunday's meals as well.

Rest on Sunday

Resting on Sunday was vital. They kept the commandment "Remember the Sabbath and keep it holy" on this day. That meant no cleaning, cooking, or hard work.

Though all these chores had their own days, the women would also prepare meals. They would mind and teach their children. They swept and dusted their home, stoked the fire, and washed dishes every day.

Women would also have other daily jobs. They would help their husbands in their businesses. Some would chop wood. Others would care for the animals, make salves and soap, weave or sew new garments, and so much more.

Our pioneer ancestors had a lot on their plate without any of the mechanical helpers we have today. But God was with them. Next time you find yourself complaining about the laundry or the dishes, count your blessings!

{Editor's note: Our family celebrates the seventh-day Sabbath, therefore our day of rest is on Saturday instead of Sunday. Juggling days can depend on music lessons, which might be a good day to go to the market or many other unique situations. Each family must decide how to apply these old-fashioned techniques to their homemaking.}

Once upon a time, prepping was NOT an option

By Kathryn White

Once upon a time, prepping was NOT an option. It was just something you did.

Great-grandma canned enough food - fruit, veggies, eggs, meat, etc. - to last 4 months because if she didn't, they would starve. Great-grandpa killed a pig every Fall and smoked all the meat because if he didn't, they would starve. Great-grandma couldn't rely on electricity to keep the house and the children warm during the winter. She had to plan and prepare to make sure they didn't freeze. Great-grandpa couldn't walk to the store to buy a cord of wood. He had to chop it himself and store it before the wood got wet.

My own grandpa once told me about how, as a boy, he had to prepare enough food and water for one day of plowing a field 2 miles from the house.

There was no car, no tractor, no cell phones, and no one living nearby who could help him in an emergency. He was on his own and had to prepare accordingly.

Back in the day, you were a prepper, or you suffered and died. Period. Today, if you are a "prepper" you might be given a tin foil hat. Wear it proudly!

I've been teased for something as simple as bringing a spare jug of water in my car on a short road trip in the summer. Heaven forbid if I had shown the person teasing me that I had a bag with a shovel, spare clothes, menstrual pads, bandages, light, and a blanket that I keep in my car at all times!

Anyone with a pair of eyes on world events can see the wisdom in staying prepared. We don't even have to look at Venezuela and Brazil and other places of chaos and upheaval. We can look at the United States.

This winter, in December 2022, a woman froze in her car in Buffalo New York on her commute home from work when she was caught unexpectedly in a blizzard that everyone was underprepared for (Anndel Taylor, NYpost.com). I can't help but wonder if she'd had a way to shovel snow away from her car exhaust and a spare blanket with her, if the story might have turned out differently.

I've known people who accidentally locked their keys and cell phone in their car and had to walk miles of smoldering summer highway in cute flip-flops that blistered their feet before they could get someone to help them.

I myself have been caught off guard by a tornado in my pajama pants on a "short" road trip that then turned into a 24-hour stay at someone's basement with no spare menstrual pads, a huge blood stain on my rear, and having to hike to my car in sandals through 3 inches of mud. It was a bad day. Needless to say, I learned to be more prepared after that.

Being a prepper is certainly not a foolish thing to do, even if you only live a few blocks away from the grocery store. It doesn't mean that you need to have a bomb shelter and a year of food stored in a hole somewhere. You don't have to wear a tin hat and believe in Bigfoot to be a prepper, either.

Having enough food and clean water available to last you a couple of weeks, and alternative ways to keep yourself warm or cool without electricity will have you comfortably sitting at home during a power outage or town-wide emergency (covid lockdown, anyone? Remember that?). Keeping some basic necessities in your car will keep you comfortable during a short-lived emergency, like spending the night in a tornado shelter. Making a habit of dressing yourself comfortably, but practically, will also keep blisters off your feet (and your pride) in case something goes sideways and you find yourself dependent on someone else to help you.

Kathryn White is native to Oklahoma, the eldest of 6 kids, and was raised on a farm. She enjoys traveling and exploring with her beloved husband, Steven, and sharing their videos on YouTube: @okiedokieexplorers When she's not traveling, she is at home, writing books or walking her dog and toddler around the block. Or maybe watching Star Trek TNG. You can keep up with her at her website: www.KathrynJFogleman.com

Preparing Your Garden Beds for Spring

By Christayla Vassar

Who can wait for those first shoots to break through the soil and green plants to take over the garden again? I know I can't! But before we can see our plants grow, we need to prepare our garden beds for spring.

Preparing Old Beds

If you're like me, you didn't do much with your garden bed when that first frost wiped out your tomatoes, sweet potatoes, and okra. Leaves piled on top of that and it's been sitting fallow all winter.

Before we get to it, you can do one optional step: a soil test. Dig down in your garden bed. Take your soil sample from 4-6 inches deep and place it in a bag. Your local extension office can help you submit your soil samples to the state university soil testing program.

With soil testing, you'll know what nutrients your garden beds possess and what they lack. Then you can add soil amendments to the garden to provide those nutrients.

To prepare your old beds for another year, you'll need:
- Garden shears
- Garden gloves
- Organic compost or good quality topsoil and humus/manure.
- Worm castings (optional)
- Straw or wood chip mulch

To start, trim all your dead plants as close to the ground as possible. You can even go into the soil, but be sure to clean your shears afterward as dirty tools can easily dull. Put these dead plants into your compost as long as they are not diseased.

But why leave the roots? The roots of your old plants build soil matter, making your soil more fertile.

Next, you probably think it's time to clean up those leaves.
No! We're going to leave them there. Leaves will decompose into leafmould, a nutrient-rich fertilizer for your plants.

Instead, you'll need organic compost. This can be homemade or store-bought.
 It's important to buy organic compost. Non-organic compost can be contaminated with grazon (same as humus/manure). If you aren't using compost, you'll need good-quality topsoil.

Please be careful how you source it! I've had family members who bought cheap topsoil from Walmart. Bermuda grass infiltrated their gardens. Last year, I liked the quality of Scott's Topsoil.

Add your compost to the garden bed. If using topsoil and humus, mix bags 4:1 and spread them over your garden bed(s). Aim for one inch deep, though up to three inches works well.

After this, add worm castings. These work much like those slow-release fertilizer pellets but without the chemicals!

Lastly, place straw or wood chip bark mulch on top. Or, if you're on a tight budget, you can also use leaves. All will protect the soil from washing away and keep out unwanted plantings. They also warm the soil up for your spring plantings. Water this every day for a few days so the wind doesn't take your mulch away!

That's it. You're finished. YOU just prepped your garden bed for spring. Celebrate it!

For more information about the benefits of no-till gardening, check this site:
HomesteadAndChill.com/no-till-gardening-benefits.

If you have any questions, you can send your emails to: thefolksysideoflife@gmail.com.

The Importance of Beef Labeling

By Andrea Hutchison

Many may say beef labeling is murky, I'd say it's dark in an evil sort of way. Ag media, "trusted" cattle and agricultural associations, Google, and even the latest AI apps harmonize misinformation abundantly "nothing to see here, buy it, eat it, and move along". Cornering truthful beef origin facts is nearly impossible. Global elites are selling the narrative that bugs are good and eating beef is bad while terms such as "sustainable beef" unleash standards that create an uneven playing field for the American cattle producer. All deployed to market misinformation to consumers, strip cattle producers' profits and eventually control people.

Trust was never a concern for our ancestors when it came to purchasing beef, they knew the farmer who raised the animal, perhaps the animal was their own, they knew the butcher who slaughtered it and they knew the grocer. They understood the process; where their food came from wasn't in question. Today people lack that important piece to the food production puzzle. Most are three to four generations removed from the farm, and its origin, making them easy targets to deceive and manipulate.

Food production and the Bible go hand in hand. In Genesis, God declared cattle good. He charged man with taking care of them and the land. Cattle utilize nonproductive areas. They are nurtured on this flawed land. It's God's design.

Along with not knowing where food originates, many today know very little about God's design. They no longer rely on scripture and have lost touch with His word. Throughout history, the manipulation of food resources has been used to control the masses. The understanding of God's production and distribution system is pretty important to a free country, it determines its survival.

Harmony exists within the diverse segments of the beef cattle production system. These functions cannot be isolated as they are necessary to achieve high-quality beef for the consumer. An understanding of what takes place in each step of the industry and how the various segments mesh together is important.

Basically, the beef cattle industry comprises six basic segments, some may be combined but this is a snapshot of the process. The purebred breeder maintains seed stock to provide bulls, the commercial producer provides feeder calves and yearlings to the stocker/background operator who in turn furnishes the cattle feeder with finished fat cattle ready for slaughter. The beef packer slaughters the cattle and provides the retailer with a finished carcass. The retailer cuts, trims, and packages the product for consumers. Interdependence exists among these segments because each one affects the cost of production or desirability of the beef product or both. The profits that each segment experiences depend on the diligence of each to perfect their efficiency in production and increase carcass quality. It is a win-win when each sector succeeds.

All sectors are connected to the land. Today threats to private property rights are entangled in beef labeling. We have witnessed the purchasing of large swaths of land by global elites and foreign countries who do not have our best interests in mind. Among other problems that face a secure beef/food supply is the recent destruction of food processing plants.

The Stockyards Protection Act of 1921 and the Livestock Slaughter Act of 1958 helped set standards pertaining to humane animal care and slaughtering. While these congressional acts were intended for good, like so many federal policies, they have been used to inject more regulations within the system.

The U.S. remains the touchstone of a quality safe beef supply. While the U.S. cattle producer is required to follow strict regulations, countries that do not are allowed to market their beef as if they are under the same guidelines. four remaining global meat packers have been allowed to place "Product of U.S." labels on imported beef over 23 different countries. Today the American cattle producer is unable to identify his product which is the safest, highest quality in the world, from countries which lack U.S. food safety standards and high quality. This has resulted in a lost competitive market. Identity in the form of a true U.S. beef label is paramount to the survival of our rural communities. Cattle and the meat and by-products they provide spawned America's rural communities. By eliminating the cattle industry we eliminate local businesses, banks, hospitals, and schools. Competition creates capital.

Cattle producers from across the U.S. cannot compete with foreign beef that has been given our stamp of approval by the USDA and FDA. Consumers are being deceived and America's cattle producers are incrementally being driven out of business. Grassroots cattle producers and cattle industry groups that understand this unfair identity theft are working continuously to protect against the threats that not only affect the producer but all humanity. Attempts have been underway for many years to remedy this deception through a MCOOL or Mandatory Country of Origin Labeling law which would force 'truth of origin' where beef was born, raised, and harvested. Currently, efforts for the American Beef Labeling Act are underway. For more information visit Label Our Beef, visit R-calfUSA.com/label-our-beef.

Andrea Hutchison is the Chain Ranch Lady of Canton, OK. She loves gathering cattle, horseback riding, helping her husband on the farm, cooking, and feeding her ranchman. She has found herself thrust into the world of politics, public speaking, and lobbying to defend agriculture businesses and protect America's food supply. Rising to the challenges before her, Andrea blogs her frustrations and victories at ChainRanchLady.blogspot.com, encouraging others along the way. Find more at ChainRanchLady.blogspot.com.

Need Seeds? Where to Find Them!

By Christayla Vassar

It's springtime and that means it's time to get growing! But where do you find the seeds to grow? Lucky for you, options abound!

Check Your Stash

If you gardened last year, you might find seeds you saved. Check where you put your collected seeds; aka, your stash. Did you put them in a basket, cabinet, or drawer? Be sure to check your refrigerator and freezer too.

Feed and Farm Stores

Feed and farm stores carry seeds. You can find organic and non-organic options. Oftentimes the stores provide a good variety to select from. They often carry Ferry-Morse and Burpee Seeds, but may also carry other brands. Tractor Supply, for instance, carries other varieties such as Seeds of Change.

Stores

Stores like Walmart, Publix, and dollar stores also carry seeds. They may carry Burpee and Ferry-Morse. They also carry cheap seed packets like American Seed Company.

Organic Produce

You can find seeds by saving them from the organic produce you buy. Conventionally grown produce may sprout, but won't always produce. Buy organic and buy local if possible.

Neighbors, Friends, and Family

Check with your neighbors! You can find some great varieties grown over several years. Plus, they are probably open-pollinated. Friends, family, church family, and co-workers may also save seeds and they may be willing to give you some. You never know who else loves to grow!

Seed Swaps

Seed swaps pop up every spring. You can find them with a quick Google search, via Eventbrite, Craigslist, Facebook Events, and even in the newspaper! You may also find seed swap groups on Facebook or Telegram with whom you can trade all year round.

Etsy

You can find some great independent seed growers on Etsy. If you're looking for grown for years or rare varieties, look here!

Seed Companies

Many seed companies exist to help you procure seeds for the growing season. Some favorites include:

- **Baker's Creek Heirloom Seeds** at RareSeeds.com - With free shipping over the continental USA, what can't you love? Some people reported problems with sprouting these last few years, but Baker's Creek offers a warranty for two years after the purchase date. They also provide reviews on each seed page to help you decide.
- **Seed Savers Exchange** at SeedSavers.org - Seed Savers Exchange not only sells its own seeds! They also allow you to buy seeds from other independent growers! Some are even free. It's like a huge seed swap. (Other independent growers can be found here: Exchange.SeedSavers.org)
- **Sow True Seed** at SowTrueSeed.com - Sow True Seed has a beautiful little catalog and great emails throughout the year to help you grow! They are also open-pollinated.
- **Southern Seed Exchange** at SouthernExposure.com - This is a great stop for all my southern friends! Southern Seed Exchange growers grow their seed in the south. So, it's already partially acclimated to the climate.
- **Fedco** at FedcoSeeds.com - Fedco sells an excellent array of seeds, plants, and trees! They send you huge catalogs of seeds, bulbs, and trees.

Others:

- **Annie's Heirloom Seeds** - AnniesHeirloomSeeds.com
- **Johnny's Selected Seeds** - JohnnySeeds.com/catalog-request
- **Gurney's Seed and Nursery Co.** - Gurneys.com/catalog_request
- **Burpee Seeds and Plants** - Burpee.com
- **Select Seed Seeds and Plants** - SelectSeeds.com/CatalogRequest.aspx
- **Pinetree Garden Seeds** - SuperSeeds.com/pages/catalog-request-form
- **Botanical Interests** - Botanicalinterests.com/catalog_request
- **Urban Farmer** - UFSeeds.com/catalog-request.html
- **Territorial Seed Company** - TerritorialSeed.com/pages/catalog-form
- **Jung Seed** - JungSeed.com/catalog_request

Find some seeds and get to planting!

Just the Facts - *Politics from the Horses' mouths...*

By Dawnita Fogleman

"RFID by 2023" has been the USDA's slogan since 2017, according to R-CALF USA CEO Bill Bullard in a January 4 2023 press release.

"The USDA implemented a strategy to lay a foundation for mandating the exclusive use of RFID ear tags whenever a cattle producer chooses to ship adult cattle across state lines," Bullard stated. "This strategy entailed the establishment and use of a private committee to develop a more detailed plan for an RFID mandate."

After some lawsuit shenanigans, where R-Calf sued the USDA and the USDA convinced the court to dismiss, R-Calf was able to keep part of the case open based on the fact that a private committee for this purpose is unlawful.

"But the government never quits when it's single-minded about implementing liberty-infringing policy," Bullard stated. "Three days after Christmas, we received a chunk of coal in our stocking."

According to Bullard, the USDA sent a proposed rule to mandate RFID to the White House Office of Management and Budget (OMB) in March of 2022 for final approval to get the go-ahead to start the formal rulemaking process.

"We should contact each of our members of Congress to urge them to swiftly enact legislation to effectively rein-in the USDA – to stop the USDA from mandating RFID," Bullard said. "Don't underestimate the importance of your calls."

Bullard encourages, it only takes one representative to understand the infringement on the rights and liberties of independent producers to be a champion in building congressional opposition.
In other news:

House of Representatives. House Resolution - H.R. 26, the Born-Alive Abortion Survivors Protection Act requires appropriate medical care for children who survive abortion procedures and imposes strong criminal penalties for failure to provide such care. Additionally, the bill protects women upon whom abortions have been performed from prosecution.

U.S. House of Representatives Concurrent Resolution 3 condemns violence that has been perpetuated against pro-life facilities, groups, and churches in the aftermath of the Dobbs V. Jackson Women's Health Organization decision.

U.S. Senators Lead Efforts to Stop Defense Department's Costly Green New Deal Mandates

Senators strongly oppose the rule and outline concerns with the proposal, including the significant regulatory burden in requiring a company to report not only its own emissions but emissions that occur elsewhere; increased costs resulting in budget inefficiencies at DOD; and the potential use of environmental reports in awarding future contracts.

https://www.congress.gov/bill/118th-congress/house-bill/26
https://www.congress.gov/bill/118th-congress/house-concurrent-resolution/3
https://www.lankford.senate.gov/news/press-releases/lankford-hoeven-colleagues-lead-efforts-to-stop-defense-departments-costly-green-new-deal-mandates

My first beaver trapping experience

Wyatt Tate, Youth Contributor

People have trapped in Potter County, Pennsylvania for centuries but for me and my dad (Brad Tate), 2023 was our first trapping season.

One of the animals we wanted to trap was a beaver.

We set out to find the beaver pond on some state gameland.

We looked for fresh chewed-off sticks, a hut, and a dam – classic signs that a beaver is in the area.

Once my dad and I found a good spot we started to explore the area and we found some beech trees that had been chewed down and sticks that were indeed freshly chewed, as well as older chewed-up trees.

We found a plot of chewed sticks where the beaver sat to feed and we knew that was an active beaver pond.

Beaver can live in two different types of homes: huts built free-standing in a pond, or a bank dwelling that is burrowed into the side of a stream.

In this pond, we found a beaver hut.

With our waders and rubber boots, we explored more to see if there were fresh slides and chewed-off sticks by the beaver hut. We were excited to see that there was a fresh sign of beaver!

We decided this was the place to set beaver traps.

The first few days we checked the traps and didn't have anything. Dad and I both had our bobcat tags filled, the first thing either one of us had caught so I thankful, but still a little sad that we didn't have a beaver.

A couple of days after that we rode with our friend John who has lots of trapping experience with beavers. We explored this same pond again and he told us where to move our traps for a better chance to catch a beaver. I was hopeful again that we would be successful.

The very next day my dad and I were checking traps and as we walked toward the beaver trap, we noticed something was different......

As we approached closer and closer, we noticed all the sticks around the trap had been chewed up! We started to get excited, could this be it?!

Dad said, "Look, Bud, we have a beaver!"

I jumped up and down, I was super excited to catch our first beaver together. Dad and I took

pictures, dispatched the beaver, and then took more pictures.

Most people don't know but parts of the beaver are edible and with the right processing and cooking are an excellent source of protein. I didn't even know that until we started trapping. I'm so thankful that God has created the beaver that I could harvest with my dad.

This is a memory I will never forget.

Wyatt is a budding outdoorsman who calls the hills of central and northern Pennsylvania home. When he's not doing his homeschool studies he is working on his Eagle Scout, and spending time outside. Follow Wyatt's outdoor adventures at wyatttateoutdoors.com and on YouTube: Wyatt Tate Outdoors.

The Homestead Act and some Shenanigans of the 37th Congress

By Dawnita Fogleman

On May 20, 1862, President Abraham Lincoln signed the Homestead Act.

For settlers to file a claim on a 160-acre plot of land, they had to pay a $10 fee and reside on the land continually for five years before obtaining ownership.

Homesteaders also had the option to purchase the land from the government after six months of residency for $1.25 per acre for 160 acres or less than 80 acres for $2.50 per acre.

For the "woke" who may read this, the provision was specified for, "That any person (specified later as he OR SHE) who is the head of a family, or who has arrived at the age of twenty-one years, and is a citizen of the United States, or who shall have filed his declaration of intention to become such, as required by the naturalization laws of the United States and who has never borne arms against the United States Government or given aid and comfort to its enemies…"

The application or claim was filed with an affidavit to the register of the local land office.

After the five-year residency, a certificate or patent would be issued to the person, widow, heirs, or devisee. If both parents died leaving children, the guardian could not sell the land for two years after. The act explicitly states it could only be sold for the benefit of infants and no other purpose.

Local land office registrars would keep a register book, tract books, and plats of all the entries and return a copy to the General Land Office with proof.

No debt could be taken on the land until the patent was fulfilled. If the land was abandoned before the certificate was received, it would revert back to the government.

The Act also gave the Secretary of the Treasury authority to classify temporary clerks with $7,000 for necessary furniture, stationery, and labor for the land office. The annual salary was set between $1,200 and $1,800 for clerks up to $3,000 per office.

I found this part of the records of the 37th Congress right after the Homestead Act to be rather funny. In the first meeting of the Levy Court in Washington D.C., after the act was passed, they were to "appoint seven intelligent inhabitants of the said county..." to be Commissioners of Primary Schools.

It then goes into the creation of school districts for the "more general convenience of the people." This in turn set up a county collector and treasurer of school funds who were not to be paid more than $100 a year.

Annual meetings were to decide a suitable and central site to lease, rent or purchase for a school-house, to vote a tax on property "owned by white persons" for building, upkeep, fuel, books, stationery, furniture, board-certified teacher's salary, and other necessary expenses.

The act is very specific in that only white landowners were to be taxed, but I saw nothing about the school being closed to any particular children.

The school tax regulations go on for many pages...

Recipe Corner

Biscuits

2 cups flour

3 teaspoons baking powder

1 teaspoon salt

1/3 cup fat, shortening or butter

1 cup milk

Preheat oven to 450°. Mix flour, baking powder and salt. Cut in fat until crumbly then add milk.

Drop with a spoon or ice cream scoop about 1 in. apart on an ungreased baking sheet. Bake 400F until golden brown, about 10-12 minutes.

Gravy

3 tablespoons fat, butter or drippings

¼ cup flour

3 cups milk

¼ teaspoon salt

⅛ teaspoon pepper

Melt butter in a skillet on medium heat. Add flour and whisk until turning brown. Slowly add milk, stirring constantly until smooth and bubbly. Turn heat down, add salt and pepper, and stir for 2-3 minutes until desired thickness. Remove from heat and add browned meat as desired.

Cajun Alfredo

4 pieces of bacon, cubed

1 lb hamburger or sausage

2 cups spiral noodles

½ onion, diced

2 cloves garlic (or more!)

1.5 cups shredded Parmesan cheese

2-3 handfuls baby spinach

½ Tbls crushed red pepper (more if you like spicy, less if you don't!)

2 pints or more of Heavy Whipping Cream or Half-n-half

Cook the noodles and butter them. Brown the bacon and onions. Brown the hamburger and garlic. Combine the hamburger and bacon mixture together. Add red pepper and spinach. Pour the heavy whipping cream til all ingredients are covered. Bring to a simmer until the spinach is soft. Add the parmesan cheese and noodles. Reheat till gooey. Enjoy!

Breakfast Alambre

(Cook in an oven-friendly skillet or, if camping, cook in a Dutch oven) Preheat oven to 400°

½ onion, diced

2 TBLS butter

3-6 slices of bacon, cubed

1 lb sausage or seasoned hamburger

2-3 potatoes, diced

1 jalapeño, diced

1 bell pepper, diced

6-8 eggs

¼ - ½ cup shredded cheese

Caramelize onions in butter. Cook bacon, but leave it a little rubbery. Add sausage and brown lightly. Drain grease if needed. Cook potatoes, browning slightly, then add peppers. Cook until soft, about 7-8 minutes. Stir, then use a spoon to form 6-8 "nests". Crack eggs into the nests. Put into the oven (or cover the Dutch Oven and put coals on top) for 5-7 minutes, or until eggs are almost done. Remove, add cheese, and melt for 1 minute.

Weather and Climate Shenanigans

By Dawnita Fogleman

With websites like WeatherModification.com available for the world to see, I'm continually amazed when someone attempts to argue with me about geoengineering. We have a stack of my husband's children's books about the weather that explain how weather modification works. He's a 1957 model, so this is not a new concept!

Archive.org has a 1978 United States Senate Report on Weather Modification: programs, problems, policy, and potential by the Committee on Commerce, Science, and Transportation as well as several videos and other documentation available.
(https://archive.org/search.php?query=geoengineering)

That particular report states, "While weather modification projects have been operational for nearly 25 years and have been shown to have significant potential for preventing, diverting, moderating, or ameliorating the adverse effects of such weather-related disasters and hazards, I am greatly concerned regarding the lack of a coordinated Federal weather modification policy and a coordinated and comprehensive program for weather modification research and development."

Let's do the math real quick. 1978 minus 25 years equals... yep, 1953. And, of course, that's just what's being admitted in this report. Who knows how long the "projects" had been going on in a pre-operational status?

I could quote all of our local politicians, "training operations" after sending them screenshots of patterns on FlightRadar24.com.

The previously mentioned weather modification website states the purposes of "cloud seeding." Well, if cloud seeding were what was happening here for the past 15 years, they missed the mark and should know by now they've potentially been the cause of extreme and even "flash" drought conditions.

Nothing takes Him by surprise.

"These deficiencies in our Federal organizational structure have resulted in a less than optimal return on our investments in weather modification activities and a failure, with few exceptions, to recognize that much additional research and development needs to be carried out before weather modification becomes a truly operational tool," the Senate Report states. "This report should include a review of the history and existing status of weather modification knowledge and technology; the legislative history of existing and proposed domestic legislation concerning weather modification; socio-economic and legal problems presented by weather modification activities; a review and analysis of the existing local, State, Federal, and international weather modification organizational structure; international implications of weather modification activities: and a review and discussion of alternative U.S. and international weather modification policies and research and development programs."

While I know there have been hearings since 1978 on the issue, the decision to promote climate change and overpopulation propaganda won the stage.

Now that the experimental mRNA has been released, we can only speculate on what is actually "seeding" above our heads. Engineering the geology of God's creation is going way beyond terra firma.

No matter what is going on, we can take comfort in God's word, "there is no new thing under the sun." (Ecclesiastes 1:9)

Nothing takes Him by surprise.

"And you hath he quickened, who were dead in trespasses and sins; Wherein in time past ye walked according to the course of this world, according to the prince of the power of the air, the spirit that now worketh in the children of disobedience: Among whom also we all had our conversation in times past in the lusts of our flesh, fulfilling the desires of the flesh and of the mind; and were by nature the children of wrath, even as others. " Ephesians 2:1-3

How Jesus Stretched Me To See the Truth About Yoga

By Tresa Salters

> *"Do not conform to the pattern of this world, but be transformed by the renewing of your mind. Then you will be able to test and approve what God's will is—His good, pleasing and perfect will."* *Romans 12:2*

I have been wrestling with this for a few years now. And though I made the decision to walk away from teaching yoga many months ago, it has taken me some time to share this.

An Alternative to yoga

After much prayer, I decided it was time to walk away from yoga. However, I love to guide people in moving their bodies to be healthy and strong. I have learned so much about the anatomy of the body, how to keep students safe during stretching and low-impact movement, and the importance of proper breathing during exercise. I have taken all this and created Prayer Moves Fitness classes.

Prayer Moves is an alternative to yoga. These fitness classes combine low-impact movement and stretching with prayer and meditation on God's word.

Here is my story

I found the exercise of yoga to be so good for my body and my physical health. It is good for the bones and for someone diagnosed with osteoporosis in my 40s, it has made a big difference for me. I found the benefits of the exercise to be so good that I trained to be a yoga teacher. I thought of it as exercise and opted out of the Reiki and some of the other new-age practices that didn't align with my faith.

After I became certified to teach, I began hearing things about how yoga was not something Christians should do. When I found Christian yoga, I thought I had the answer. Looking back now, I know that Jesus is the only answer and this was yet again leading me astray, let me explain more.

Christian yoga is an oxymoron

I enrolled in a Christian yoga teacher training and then another one, thinking this would work out great. The interesting thing is that those trainings actually opened my eyes even more to the truth about yoga. Jesus really used this time to stretch me, bring me closer to Him, and teach me the truth about this practice. I began to question it even more.

Most people who practice yoga, and many who teach as well, have no idea that the roots of this spiritual practice are so far away from our Christian faith. It really is a spiritual practice in and of itself, deeply rooted in pagan principles and opens the door to new-age beliefs. Even if you practice as "just exercise" as many do, it brings you to a spirituality that is disconnected from Christ.

> *"Jesus answered, I am the way and the truth and the life. No one comes to the Father except through me." John 14:6*

Caring for the body God gave you

The exercises involved in yoga are not themselves a problem, you can only move your body in so many ways. Stretching is good for you and many of these stretches we see done in other exercise programs and used by athletes to warm up and stretch their bodies. I tried to justify it by praying through it, but my research showed me that many of the poses done in yoga have meanings deeply rooted in pagan practices. The sun salutation, as it is called, is meant to worship the sun and its gods. As Christians, we know that there is only one God and we must not worship any others.

> *"You shall have no other gods before me." Deuteronomy 5:7*

Focus on Jesus, Don't Empty Your Mind

Meditation is also a big part of the practice of yoga. They instruct you to empty your mind, something that I have never been comfortable with. I am grateful for that because it led me to question more about yoga and realize that God does not want us to empty our minds, but rather to calm the mind and focus on Him. I began to take time to meditate on God's word and His truth and it gave me all the answers I was searching for.

> *"Oh, how I love your law! I meditate on it all day long." Psalm 119:97*

The more I read God's word, the clearer it became to me that teaching fitness was okay but teaching yoga was not. God calls those who are teachers to a higher standard, and I wouldn't want to lead anyone astray doing a yoga practice. The verse from 1 John 4:1 kept coming up for me over and over as I was researching yoga and other new age practices, which I will not get into here, but I will warn you that there are many other things that open doors to a way that is not of God and you must be careful to discern what you are getting involved in:

"Dear friends, do not believe every spirit, but test the spirits to see whether they are from God, because many false prophets have gone out into the world." 1 John 4:1

God will guide you on the right path

After much prayer and listening to God's guidance, I stopped teaching yoga classes. It is important for us to care for the body that God gave us and fitness is a big part of that. There are so many benefits including strong bones and muscles, keeping your blood sugar under control, and a healthy mind. I have over 600 hours of training in the anatomy of the body, how to keep students safe during stretching and low-impact movement, and the importance of using the breath properly during exercise. I took all this knowledge and created Prayer Moves Fitness classes.

Prayer Moves is an alternative to yoga. These fitness classes combine low-impact movement and stretching as we pray and focus on God's word throughout the class. Classes are online, though I do plan to begin offering in-person classes in early 2023, we also join together for monthly prayer calls. You can learn more here: Prayer Moves Fitness program - https://mailchi.mp/53f05b43f9e3/prayermoves

There are many great resources that I have used to seek the truth about yoga. I am grateful to so many who have had the courage to come forward and share how they wrestled with this decision and the path that led them to the truth. You will find many videos on YouTube and articles with a quick search on the internet.

This website was very helpful to me: Truth Behind Yoga (TruthBehindYoga.com)

May God bless you on your journey to care for the body He gave you! My goal is to stay in His truth as I help you to live your best life! I would love to stay connected with you, you can join my email list for weekly inspiration and health tips.

Tresa Salters is a Christian Health coach, who helps women solve stress and overwhelm through healthy lifestyle choices for their body, mind, and spirit. By creating healthy habits that will last, Tresa leads students through "Prayer Moves" fitness classes for Christian women. She has over 600 hours of fitness and anatomy training and leads low-impact stretching and movement classes both in person and online. These classes are faith-based, praying and meditating on scripture while healing the body and calming the mind. You can find Tresa at LiveWellBlessed.com or email blessedtres@comcast.net.

101 Ways to Kill Sourdough

By Dawnita Fogleman

I'm a homesteading failure in a LOT of ways. There are so many times I've failed utterly and totally in my endeavors to get back to the basics. The trick is to not let these failures get you down. It's important to "fail forward," growing and learning along the way. Perseverance and determination are homesteading virtues! While we definitely DON'T have to do ALL the things, don't give up on something you really want to learn. Keep trying.

1. Give it BLEACHED flour.
2. Forget to feed it for a few months.
3. Put hot water in it.
4. Let it get too cold.
5. Use chlorinated water.
6. Keep it next to other ferments
7. Keep it too warm
8. Not feeling it often enough
9. Not discarding it when you are starting a brand new starter
10. LOL... you get the idea!

Thankfully, Traditional Cooking School has wonderful tutorials to help with so many basic, old-fashioned homemaking skills. One is the How To Start A Sourdough Starter. This is a free resource (with my affiliate link): TradCookSchool.affiliatedash.com/a/starter/a7976

In 5 minutes, you can mix up your sourdough starter and set it on your kitchen counter to do its thing. Plus, taking a HUGE first step toward healthy bread for the family! Get the free & simple instructions today.

Fostering Community

Spring 2023   PrairiDustTrail.com   Volume 01 Issue 02

PrairieDustTrail.com
Connect with the past ~ Prepare for the future ~ Live more sustainably

Fostering Community

- Starting Seeds
- Poultry Care
- Practical Rabbit Care Tips and Tricks

Editor's Epistle

By Dawnita Fogleman

This spring, it seems like everyone got all the April showers except Northwest Oklahoma. We're sitting in a pile of blowing dust. My hopes for a beautiful garden have diminished to desperately praying for rain. We have started some seeds, we'll just call them seeds of hope for now. On a positive note, the top is on the greenhouse. We'll be stocking the tower garden in there soon and harvesting goodies long before the outdoor produce is ready. Keeping things cool will be the trick once summer actually sets in. There are always struggles. Keeping the critters out of the garden, braving the weather, and overcoming obstacles along the way are but a few.

As we publish this edition, I'm finding myself struggling with the theme. I admit "Fostering Community" has not been my top priority.

Humans tend to take each other for granted, even in the best of circumstances. A few times burned really takes the enthusiasm out of participating in community activities.

Since social media, it's been even harder to really connect and have relationships with local people. The more conveniences we have, the less time we have. Have you noticed that?

I'm excited about what our contributors have shared though. I've been convicted and encouraged to reach out more in our community. I hope you are encouraged as well!

Forging kinship through struggle and differences

By Pastor Ken McKinley

Communities are built out of relationships, and of course, being a Christian, I believe that the Bible is the best source for learning how to do that. One such passage within Scripture we might look to is 1 Timothy 5:1-8. Within that passage, Paul gives us some great practical guidelines in which to build community.

In those eight verses, he admonishes us to show respect for one another. We are to treat those within our community as family, older men as fathers, older women as mothers, and younger men and women as brothers and sisters. But this goes deeper than treating someone in a certain way, because ultimately how we treat others is largely dependent upon how we see them.

Think about this for a second; I look at my cattle dogs as helpers and companions. They help me round up goats and keep coyotes at bay, I feed and shelter them, and they bring me a lot of joy and happiness. Someone else might see them and think of them as simply pets. The coyotes look at them and see them as a hindrance to their goals (namely eating my goats).

Within a community, it is much the same. We might see our neighbors as annoying troublemakers, we might see them as fellow farmers, ranchers, and homesteaders, someone we could learn from and share experiences with, we might see them as people created in the image of God, and deserving of respect, honor, and compassion. It really does depend on how you view others.

The question then is – how should we see others within our community? In society at large, people base their views on certain aspects like wealth, power, social standing, etc. If a person seems wealthy, they tend to be treated better, whereas a person who seems poor is often not given the respect and esteem that the wealthy are given. This is true even in rural communities. If a farmer or rancher has ten thousand acres, he is often afforded more respect than the poor guy who only has a couple of hundred acres. And this is where the Christian faith plays a vital role in building communities. A particular person's wealth or lack thereof does not affect or determine their standing before God. Nor should it influence how we view that person or treat that person.

Instead, we see them as fellow human beings, created in the image of God, who is, like us, trying to make the best out of life with the hand that has been dealt them. Regardless of their wealth or social status, they have inherent worth because of that fact alone. But they also have a unique value in the sense that each of them has knowledge, experiences, and insight into certain issues that are unique to them as individuals. They all have their struggles and victories, and we can learn from them, and hopefully, they can learn from us.

And so as we begin to see our neighbors as more like family rather than just someone who lives down the road, relationships are forged, and community develops. And it's a wonderful thing.

Ken McKinley has spent his lifetime in outdoor activities, beginning in Boy Scouts. He is a husband, a father, a homesteader, and veteran of the armed forces, having served six years with the US Army 3rd Ranger Battalion. He's an ex-law enforcement officer, a black belt in Judo, and a Licensed self-defense instructor. McKinley has been a senior pastor for twenty-four years and has a Master's degree in Theology from BMATS, working towards a master's in Criminal Justice through American Military University.

Starting Seeds

By Christayla Vassar

You can't miss it. When you go to get your groceries or feed, there they are: vegetable plants for your garden, begging you to take them home.

You look at the price. Ouch! $5 for a single tomato plant?! Surely there's a better way to plant your garden on a budget.

That's where starting your own seeds comes in. It's getting later in the year, so figuring out what to start and where is crucial to do NOW.

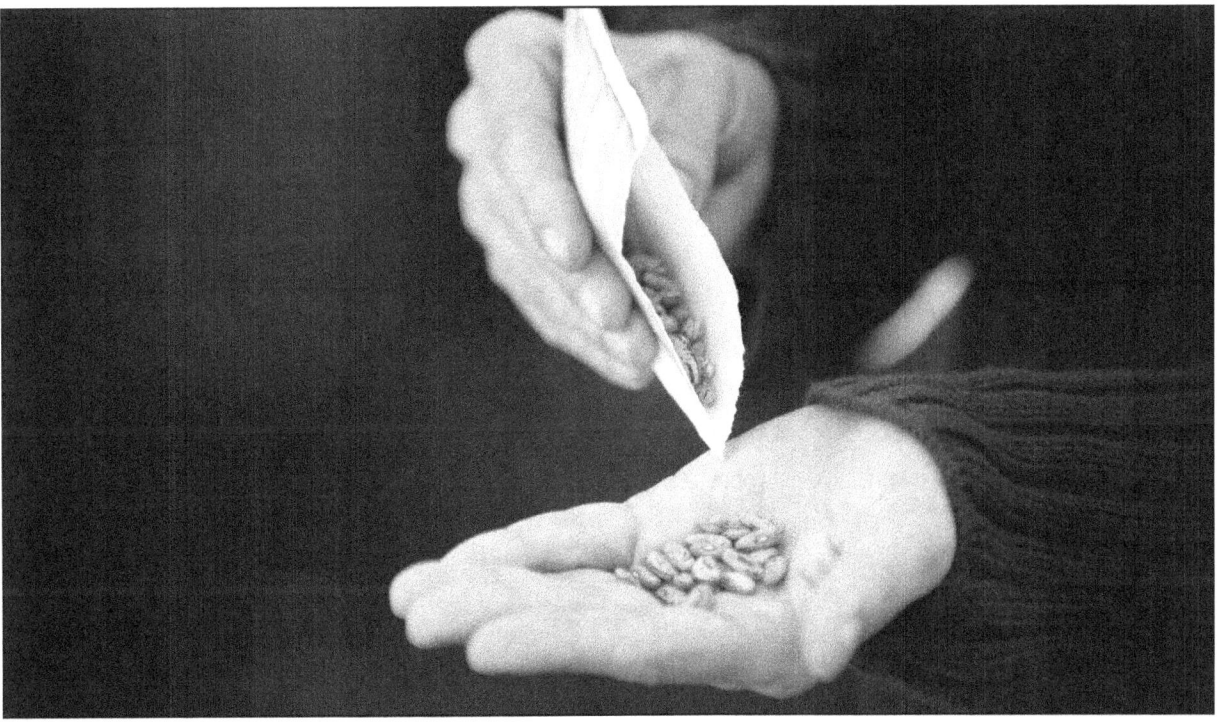

What Seeds Can I Start Indoors?

You can start most seeds indoors. Potatoes will naturally start to grow eyes in January or February in many parts of the country.

You can start sweet potatoes can by encouraging slips to grow. You can encourage sweet potato slips to grow by suspending the rooting end of a sweet potato in a jar of water. The rooting end is usually the tapering end of the sweet potato.

Other seeds to start indoors:

- Tomatoes
- Tomatillos
- Peppers
- Eggplants
- Onions

- Cabbage
- Broccoli
- Cauliflower
- Celery
- Melons

Herbs such as basil, calendula, anise hyssop, comfrey, thyme, chives, oregano, marjoram, parsley, dill, cilantro, and lemon balm.

What do I need to start plants indoors?

You'll need:

- Seed-starting soil mix (or an airy potting soil mix)
- Filtered water
- A bucket

- Small pots or seed starting tray
- A sunny windowsill or a grow light
- A spray bottle filled with filtered water

1. Place your seed-starting soil mix in a 5-gallon bucket.
2. Add filtered water and mix until your soil is no longer dry.
3. Fill your small pots with your seed starting mix and plant your seeds. Plant seeds as deep as they are tall. (For example, if you are planting tiny seeds like basil seeds, sprinkle them on the top of your soil.)
4. Place them on a sunny windowsill or under a grow light.

Water with your spray bottle daily.

How do I cold stratify?

Cold stratify your seeds to sprout by putting the seeds in a plastic bag with damp soil. Leave one corner of the bag open and check the bag to be sure the soil doesn't dry out. Add filtered water if it does. Leave the seeds in the refrigerator for 6-8 weeks, then try to sprout them normally.

What Seeds Do Better Started Outside (direct sown)?

- Okra - Okra grows a long tap root to anchor it into the soil and to mine for minerals. If the pot isn't long enough, it can damage the root and stunt plant growth. I recommend direct sowing instead.
- Root vegetables - Most root vegetables do not transplant very well. While you can start them in a sunny window or under a grow light, you must start them, in the pot they are going to be grown in. These include: carrots, turnips, beets, radishes, parsnips, potatoes, and rutabaga. Onions are the exception to this.
- Corn - Corn grows quickly, so unless you're looking to harvest earlier, it does better directly sown into the ground.
- Vining crops - Vining plants include ones such as squash, beans, peas, and cucumbers. These plants grow quickly and need room to expand.
- Lettuce and Greens - These grow so quickly and do well in cooler temperatures.

Troubleshooting

My tomato/pepper/tomatillo/eggplant plants won't sprout/grow. What can I do?
These nightshade plants do better when the soil is warm. If you are having trouble sprouting or growing these plants, place them somewhere warm. You can buy a heating mat or you can try to place them on top of your refrigerator. It sounds odd but it works!

My seeds aren't sprouting!

How old are your seeds? Seeds do not last forever and will die eventually. Did you soak them or scarify them? How deep did you plant them? Has it been up to 4-8 weeks since you planted them? Have you been keeping them moist? Have they been getting enough light? Ask yourself these questions and try to rectify the problems. If they still won't sprout, try again with new seeds

My plants have thin stems and fall over easily.

For strong stems, a plant needs plenty of sunlight and exposure to the elements. Believe it or not, the wind helps makes plants big and strong! Take them outside. Start for an hour in the morning sun on a day the wind won't break their stems or knock them over and increase the time from there.

Poultry Care

"I swear chickens are the easiest, hardest animals to raise!" ~ Kathryn White

- Apple Cider Vinegar (ACV) - Just a bit in their water. We do about ½ cup per gallon of water.
- Wormer - Ask your vet the best way to do this.
- Minerals - Liquid all-stock or block
- Colloidal Silver - This can be alternated with ACV in water.
- Is the coop clean of mites, lice, and other pests?
- Food Grade Diatomaceous Earth - Sprinkle in the coop and yard area.
- Probiotics - Some of that over gown kombucha mother works wonderfully.
- Egg Shells - You can crumble and toast them in the oven to detour them from eating their own eggs.
- Do they have sand and gravel for their gizzards?

And, of course, pray over your birds, land, and buildings.

Quail Basics and Some More!

By Francis Roland, Youth Contributor

What do you think of when you hear "quail"? Do you just think of alphabet books with a picture of a little bird next to the letter Q? Today, I'm going to teach you more than you ever thought there was to know about quail.

Quail come in all different colors and sizes. They are found all around the world. The most common breed of quail is the California quail. They are the ones with a question-mark-shaped feather on their heads.

There are quail that weigh two to three pounds, like the Courier and Bob-white. There are also ones that weigh three to five pounds like the Gambil and Californian. Most are brown but they can be white, gray, blue, tan, and black as well as many other colors.

I'm going to be talking about the Cornix quail which is one of the most common quail to raise. They are about the size of an average tennis ball and lay eggs the size of ping-pong balls.

They are very easy and affordable to raise. You can find them at your local hatchery or farm store.

Cornix quail are usually about $5 each. The feed is also easy to find. We give ours chicken grower with at least 16 % protein mixed with sunflower, corn, oat, or whatever grain we have.

You can also tractor (graze in a portable pen) the quail and feed them your kitchen scraps. The average amount of food a quail eats in its lifetime is 30 pounds. In that same amount of time, that quail can lay 750 eggs. The average cost of raising a quail for an entire life cycle is about $20.

How to Get Started with Quail

First, we should start with the question of do I want to raise quail. Quail are great for meat, eggs, money, and fun. They are easy and cheap to raise, so here are the steps to starting your own quail flock.

First, you will need fine bedding in the brooder like wood chips or chaff in a thin layer. You can use a plastic storage tote as your brooder. On top of it, you will need a heat lamp and a screen so they cannot fly out.

For the feeder, you can use a regular chick feeder. However, for your waterer, you will need a special base or need to add rocks to the bottom of a regular chick waterer to keep the quail chicks from drowning. In one storage tote, you can brood about 20 chicks.

After around two and a half to three weeks, the quail are ready to go to their coop. Their coop should have a run (a fenced-in area) that is enclosed with hardware cloth because raccoons can rip through chicken wire, and a (shelter) box made of wood or other material that blocks sunlight.

Your coop should have a minimum of one square foot per bird. You should provide them with shade and some protection from the wind but be sure to let them have plenty of sunlight or they won't lay unless you provide artificial light.

Inside your coop, you can use any feeder but to not waste as much feed, you can put a container under the feeder to catch the grain that gets knocked out. You should also give them a bowl of sawdust or dry dirt so they can take dust baths. If you set it up the right way, you can take care of feed and water once per week and just collect eggs once per day.

The Economics of Quail

Quail eggs are 2/3 yolk and about the size of a quarter. Three quail eggs equal one chicken egg. Each hen on average will lay 250 eggs per year. They begin laying at between five to seven weeks of age.

You can harvest them for meat at about five weeks, too. The meat is dark and tasty. Each quail weighs about one pound after butchering. They are very easy to process.

Fifty-pound feed bag costs about $20. You can feed five quail for one year with just one bag of feed. A quail is ready to harvest in just about one month. So over the course of a year, you can raise 60 quail for meat or eggs for the cost of one bag of feed. (12 months x 5 quail = 60 quail for one year).

Each quail will cost 33 cents to raise to maturity and you can find chicks for $3-$5 per chick. This means that it will cost you between $3.33 and $5.33 to raise one quail for meat.

If you are just looking for eggs, you can start with the same $20 bag of feed. This will feed 5 quail for an entire year. Which will cost you $4 per quail for the year. Each hen will lay about 250 eggs which means each egg will cost you only 2 cents to produce. Three quail eggs equal one chicken egg. So for $0.72, you can get the equivalent of one dozen chicken eggs. That means you will get the equivalent of 83 chicken eggs for just $4 of feed. Your total first-year costs will be $7-$9. Most chicken eggs cost around $4 per dozen at this point.

You can raise quail virtually anywhere. They need very little space and are very hardy. Quail

can survive temperatures between -20 and 100 degrees Fahrenheit without any special care. Even if you live in town or in a city, you can have them in a tiny coop in your yard, garage, or anywhere that they can get good sunlight if you are raising them for eggs. Now you know all the reasons you should raise quail.

Francis Roland is a young teen who raises quail and other animals on his family homestead in rural South Dakota. He is on a mission to have everyone raise quail for sustainability and fun.

Building Communities

By Elizabeth Santelman

Life on a farm is not for the faint of heart. There are hours of work to be done and so many people and things relying on you. It's so easy to be so busy and overwhelmed that we begin to sacrifice our own well-being and human flourishing and call that noble and good.

For many women, this looks like giving up relationships with other women. Labeling it as superficial and unnecessary gives us a sense of purpose in the discomfort, but nagging at the back of our hearts is the feeling that a piece of who we were created to be is not its full potential.

We stay late at church attempting to cram in all the connections we didn't get all week. Our husbands and children wait in the car patiently... or not so much.

What we may not realize is that human friendship isn't just a fluffy extra added to life, but a fundamental aspect of who God created us to be... Do believe me, hang in there and give me a second.

Genesis 1:26 says, "And God said, Let us make man in our image, after our likeness:" Notice what's happening there God uses the word "us" about himself. We know from more of the Bible that God isn't one but rather a trinity. God is forever in community with himself as a trinity... Still not convinced... Try this...

Genesis 2:18 says, "And the LORD God said, It is not good that the man should be alone;" After he declared this he made a woman and ensured that humans would be multiplied. Sure this means it isn't good for a man to be without a wife, but the verse doesn't specifically say that. God is declaring that as humans we need other humans.

There we are two chapters in and God establishes that people need people. That's fine and good, but if you live rurally or don't have much time how do you make friends? I would love to give you some really practical ways to make that happen!

1) Don't be picky

I bet you didn't expect that from a Christian magazine, but here we are. So often as the American church, we divide ourselves up so narrowly, and in that we miss out on the beauty and benefits of learning from and loving others different from ourselves.

I'm not asking you to share your deepest secrets with the town gossip, but maybe consider inviting someone who may have different interests and life experiences from yourself. Listen to their story without judgment. I've learned all humans are endlessly fascinating. You may end up becoming close to someone you didn't expect.

2) Start with social media.

Please, don't end your relationships there, but follow other local women. These may be ladies from your church, other churches, or neighbors. From there you can begin conversations, and see what they are interested in so those in-person meetings aren't as awkward. It gives you questions to ask and topics to explore that you know they will be interested in. Start looking for common ground.

3) Don't wait for perfect

Maybe this means you invite someone before your house is clean. Maybe this means you invite someone to come can beans with you instead of waiting for the winter when the garden season is over. Most people would rather just be invited into a mess rather than have you wait for the perfect moment, and never extend the invite. As women, our work will literally never end. If this isn't the right moment, you probably won't find one that is.

Another way you might be waiting for perfect is waiting for the perfect person. I hate to break it to you, but you're never going to find someone who thinks and believes exactly the same thing you do. If you're waiting for that you're going to be waiting a long time.

4) Make a list

Every year I like to make a list of what I call "My 5". I've found if I try to focus on too many people I end up going deep with no one. So each year I focus on building relationships with those same 5 women. Sometimes in my busy life, this looks like just sending a text message that week and checking in. Sometimes we can spend face-to-face time together, but regardless my hope is to have a closer relationship with them at the end of the year than I did at the beginning.

5) Take the first step

Often where we go wrong is by waiting for someone else to take the first step. Of the women I know who have friends, they were always the first ones to reach out to most of the people in their lives. We have this idea that it's more flattering to be asked, but really the favorite person in a group is the one who does the asking. Make others feel good.

If this scares you, try doing it with your kids the first time. Meet at a park, or neutral location rather than your home, so you have an out if it's getting awkward. Knowing there is an escape plan may help you get over the fear of asking.

So get your pen and paper out. Write down at least one person you want to try this with if you had time to read this article you have time to send them a quick text message. Bonus points if you set up a time to meet! I can't wait to see how God blesses your relationships and faith in this coming year by building the church by building communities!

Elizabeth Santelman – Is an Illinois farm girl turned Okie. She currently lives in the city, because she fell in love. She has three little boys who she is teaching to garden, and to love hard work outside (it's shockingly hard to get stuff to grow in Oklahoma). You can often find her with a cup of coffee, dirt under her fingernails, pecking away at the laptop and writing down her latest ideas. You can see more from her life and find more encouragement at SunshineinMyNest.com or @sunshineinmynest on Instagram.

5 Plants to Forage in Spring

By Christayla Vassar

Spring has finally sprung! After the cold of winter, greenery is peeking its head out again. These first plants feed the wildlife and insects, but there are several that can feed you too!

Remember to harvest from areas you know haven't been treated with pesticides or herbicides. If you are unfamiliar with an herb and how to respond to it, only drink up to 6 to 8 ounces of tea made from it and wait to see if you have a reaction.

Dandelion (Taraxacum officinale)

You've seen this beautiful yellow flower and its green leaves sticking out of your yard. Some people call it a weed, but it is actually a powerful spring-time first food in nature. The leaves can be long and lobe-like or toothed, with a yellow flower extending from a long, smooth stem.

You can eat its young greens in salads and sautéed. The older they get, the more bitter they get. Though still edible, you must cook and season them. You can use the flowers in salads, syrups, or jellies. You can also use the root, but that is better harvested in the winter. Some people roast it and use it as a coffee substitute. It's full of vitamins A, B, C, and D and iron, potassium, and zinc.

Walther Otto Müller (1833–1887)
Köhler's Medizinal-Pflanzen

proprietà medicinali, Piante, orti, Angelo Franciosi, poesie, Veneto, 1805-1822

Medicinal uses: diuretic, choleretic, anti-inflammatory. The root or leaves are the usual part taken as medicine. It's taken in tea or dried in capsules. It also detoxified the kidneys.

Henbit (Lamium amplexicaule)

Has purple taken over your yard? If so, you may have henbit. Henbit is a green that has tiered leaves with little purple flowers. The edible younger plants make great additions to salads, sandwiches, soups, smoothies, and stir-fries. It's full of vitamins A, C, E, and K and calcium, iron, manganese, and magnesium.

Medicinal uses: anti-rheumatic, anti-inflammatory, diaphoretic, febrifuge, laxative, anti-diarrheal.

Taken as tea usually. It can help with inflammation, soreness, or stiffness. It can help with relieving the unpleasantness of fevers or breaking them. It also balances the digestive system (helps with constipation or diarrhea). It calms menstrual cramps and excessive bleeding. Externally, use the tea as a wash or crushed plants as a poultice for burns, cuts, scrapes, and stings.

Dead Nettle (Lamium purpureum)

Do you have a green that looks like henbit with beautiful purple flowers but with larger, heart-shaped leaves? Does it look a little dead (purple leaves on top and slightly wilted)? Congratulations, you found dead nettle!

Add dead nettle to salads, soups, and stir-fries. It can also make beautiful pesto! It's fabulous as a dried herb for tea and salves. It contains vitamins A and C, plus iron!

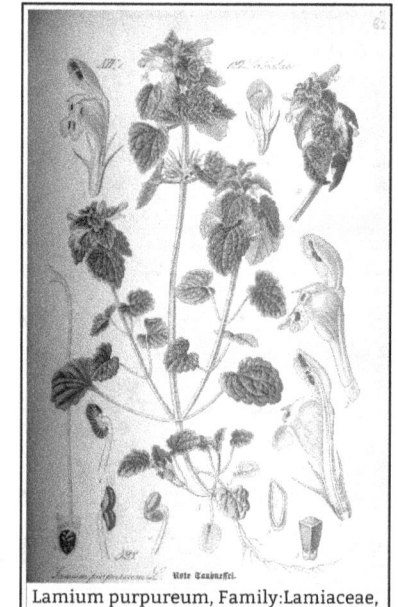

Lamium purpureum, Family:Lamiaceae, 1885,Prof. Dr. Otto Wilhelm Thomé Flora von Deutschland, Österreich und der Schweiz 1885, Gera, Germany.

Medicinal uses: antihistamine, anti-inflammatory, diaphoretic, diuretic, laxative, anti-fungal, anti-bacterial. Dead nettle makes a great tea or wash, especially for springtime allergies or skin irritation. It's also useful for fungal or bacterial infections. It can be made into a tincture or salve as well.

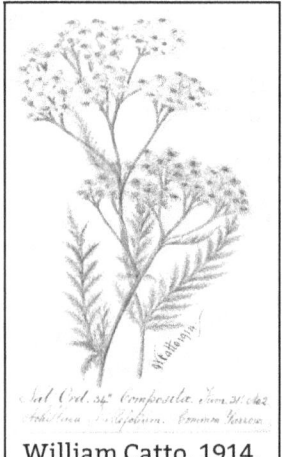

William Catto, 1914, watercolour on paper

Yarrow (Achillea millefolium)

Yarrow has feathery leaves and a tall stem holding a rather large cluster of white flowers on top. It has a distinctive smell when crushed that reminds some people of chrysanthemums, others of pine needles, and yet others of a spicy herb like rosemary.

You can use it in cooking. The leaves taste bitter and the flowers peppery, so it makes a great addition to a salad. The leaves can also be added to any dish as a vegetable. Yarrow contains vitamins A and C, plus the minerals zinc, potassium, magnesium, phosphorus, niacin, and calcium.

Medicinal uses: astringent, analgesic, styptic, anti-microbial, coagulant.
The tincture can repel mosquitoes. The herb, dried or chewed, can help stop bleeding externally or internally. When infused in oil, the herb can help with bruises, sore and inflamed muscles, or even wounds. Yarrow tea can help out with colds and minor digestive issues. It can also work as a poultice or wash to relieve pain and help injuries heal.

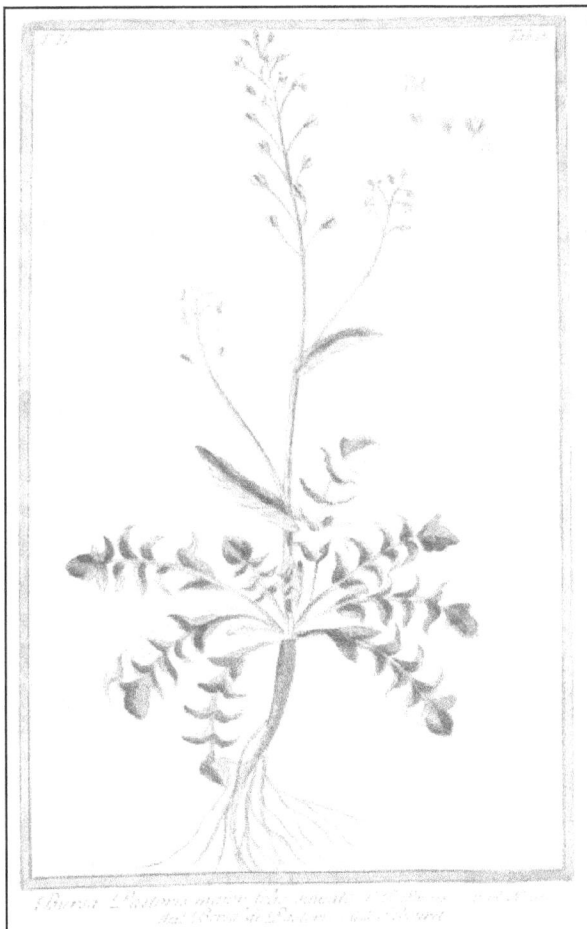

Hortus Romanus secundum systema Turnefortii a Nicolao Martellio Linneanis characteribus expositus, adjectis singularum plantarum analysi ac viribus ; species suppeditabat ac describebat Constantinus Sabbati. Engraved colored vignette[Shepherd's purse] 1772

Shepherd's Purse (Capsella bursa-pastoris)

The leaves of shepherd's purse are toothed. The stem is tall with heart-shaped leaves running up and down it. Small, white flowers crest the top in small clusters.
You can eat the leaves and flowers raw or cooked. The root is spicy like ginger and can be used in place of ginger. The plant is high in vitamin K and also contains vitamin C and iron, potassium, and calcium, Medicinal uses: stimulant, vasoconstrictor, vasodilator, coagulant, hemostatic, anti-diarrheal, laxative, anti-inflammatory, anti-fungal, diuretic
Yarrow is well known as a way of stopping bleeding internally or externally. It can be tinctured or used as a poultice on wounds. The tea can be used for internal bleeding and is used after childbirth in many places.

Happy foraging!

Practical Rabbit Care Tips and Tricks

By Prairie Ruth Fogleman

Rabbits are prone to many diseases and can die from pretty much anything for any reason. So here are a few things I do to keep them happy, safe, and healthy.

First off they need a good sized clean hutch or cage as their safe calm area with a box or hidey hole, toys (cat jingle ball toys and plastic baby keys are some favorites here) and stuff to chew on as rabbit teeth never stop growing a good pace of apple wood is always good for them.

Good quality everyday pellet feed is a must here, they can't live off of veggies, fruits, and hay alone, veggies and fruits can often cause intestinal problems from the sugars they can die from, so just a small piece every other day as a treat is best. Hay can be free-fed or a handful daily to help that gut move well.

A teaspoon of apple cider vinegar in a quart of water once a week or 3 times a month is very beneficial for any animal, it has natural products that they need especially in the spring and summertime.

Colloidal Silver is an every-other-day thing for mine, it boosts their immune system and fights off bacteria, they can drink it and or put it in their eyes. I put a quart in two gallons of water for them to drink and put a drop or two in their eyes. Really helps keep them healthy and happy, especially in this dry dusty drought.

I sometimes add probiotic powder or paste to one's food if I've got one that's really struggling with all the others listed above.

Engagement Nurtures Lasting Community

By Micheal Hein

Fostering community can be defined as, encouraging and promoting the development of common characteristics and interests within a group of people. When one thinks of community it is likely the norm to think of people who are living in close proximity. However, in our day and age that may not necessarily be the case. With the internet, communities can involve people from anywhere on the entire globe, and people who may never meet in person. There are many other considerations that must be addressed besides the possibility of Internet participation in a community, which depend upon what kind of structure and participation guidelines the group desires. There are also negative influences which work against community that should be addressed, things which could potentially cause the community to cease functioning.

If one has heard stories from people of past generations, one may have recognized that the lack of modern amenities often resulted in greater amounts of community and social interaction. These social gatherings ranged from neighbors helping each other harvest crops or work cattle or just gathering for food and fellowship. When we look at the circumstances that surrounded these types of get-togethers, one common denominator usually present was the lack of something.

In the realm of farming, the lack was most often that of the necessary equipment to complete a task, such as a threshing machine. This prompted the development of threshing crews, where a threshing machine was taken from farm to farm and neighbors would help each other thresh their wheat. These events didn't only involve the men, but also the women. Food and water for the crew were equally as important to the process and were provided by the women.

When it comes to other types of social gatherings, these were also encouraged by the lack of something. Some of those things lacking were TV and movies and other forms of entertainment. A friend of mine has related how, as a child, neighbors would come over regularly to visit and play music. However, when his family and their neighbors got TV the visits came to an end. Church socials and the community at large would also engage in more social events in times past before these other influences came into existence.

When it comes to fostering community it may be easier to get started than it is to maintain. In the beginning, there may be excitement and newness about the idea but there is always the danger of distraction. There is continual pressure that pushes to isolate people from each other and draw them away from the community. As society progresses there will inevitably be things come along that are detrimental to a community if their influence is allowed to do so. Therefore, the steps to protect against disruption and distraction may be as important as the steps to begin.

Micheal Hein farms with his sons and growing grandchildren. His family homesteaded in the Oklahoma "gateway" to the panhandle in the 1920's.

The Importance of Teaching Community to Your Children

By: Hope Ware

Over 50 years ago, I grew up in a small town, surrounded by two things: cows and cornfields. With a total population of 1800 souls, every adult knew your first name, address, and who your parents were.

When I misbehaved on the four-block walk home from school, my mother was waiting at the front door, having been informed by a neighbor of my poor decision to cut through Ms. Crabb's backyard (yet again).

When a student at the school was diagnosed with leukemia, every child from age five to eighteen was seen walking the streets while carrying small, metal cans the following weekend. They were canvassing neighbors to raise money for The St. Jude Telethon.

When a tornado tore through the west of town on a Saturday night during a showing of the latest blockbuster movie, one mother took every child at the theater to her home, keeping us safe until our parents could weave their way through darkened city streets to pick us up.

All of these stories are examples of what it means to live in a community.

When my children were young, I wanted them to understand that they were an integral part of our community and could serve, love, and comfort others within their reach.

Our family coined a special word for this. We called it, "one anothering". This phrase was born out of the Bible verse Romans 12:10:

> *"Be devoted to one another in love. Honor one another above yourselves."*

"One anothering" is best described as a natural outflowing of your love and concern for your family, friends, neighbors, and community.

The smallest community your children will experience is, of course, your nuclear family. This is followed by extended family. However, there are also multiple communities in which they will be involved: church, school, neighborhood, sports teams, and city being just some of them.

Even though children are small, their special touch is needed in the lives of others.

The importance of showing love to those around us is clearly demonstrated in scripture.

John 15:12 states, "This is my commandment, that you love one another as I have loved you.".

Simply put, being the hands and feet of Jesus means thinking practically about the feelings and needs of those around you.

The Beautiful Little Greenhouse (1810) by T.J. Beyer (archive.org)

My children grew up with Miss Nailing living next door. She was a middle-aged African American woman who had never been married. She had no children but loved mine like a grandmother.

When my oldest boys were about eight and ten years old, on snowy mornings I had them get out of bed bright and early, dress warmly, and go outside to shovel Miss Nailing's sidewalk. I explained that although Miss Nailing didn't go to work until after lunch, we wanted her to wake up to a perfectly clean sidewalk.

Weeks later I noticed that they had begun to watch for Miss Nailing to return from her weekly shopping trip. I was so pleased to see them quickly bound from the house to rush to aid her in carrying her grocery bags into her home. They did this of their own volition. I hadn't even mentioned it to them.

One day, when they returned from their task I said, "Boys I am so pleased and excited that you saw a need and helped Miss Nailing. You are really "one anothering!"

They grinned from ear to ear as my second-born son, John, replied, "Yeah, Mom and she buys great cookies too!" Well, I guess a little reward for kindness is okay sometimes, too.

Here are some more examples of how children can participate in the Biblical mandate to show love, kindness, and compassion to others.

- When you help stock the shelves of the church food bank you are "one anothering".
- When you mow our neighbor's lawn while they are on vacation you are "one anothering".
- When you get a drink of water for your thirsty sibling you are "one anothering."
- When you make soup for a sick neighbor you are "one anothering".
- When you color a picture and give it to someone who is sad you are "one anothering".

Once kids get the idea of "one anothering", they tend to generate their own, unique ideas and adopt their own causes.

One Sunday morning as our pastor announced that the church was starting a food bank, my eldest son leaned across the pew and excitedly whispered, "Mom, I want to help with that!" Our family was involved with that ministry every month for the next ten years.

Sit down with your children and ask for their input. Believe me, they will bless your heart as they tell you how they would like to practice "one anothering". They will come up with ideas that you would never have thought of or considered!

Yes, the sense of close-knit neighborliness, which was the norm in the 1970's, can (and should) still be found in this new century. It happens when we teach our children to "one another".

Hope Ware and her husband Larry, frugal-living experts, have raised four sons debt-free, including paying cash for a home when their income averaged about $40,000 a year. She teaches families practical frugality skills so they can pay off debt and live the life they love while keeping their spending under control. Join over 150,000 fans on her YouTube channel @UndertheMedian and at her website UnderTheMedian.com.

Death by a Thousand Cuts:

The Incremental Path towards Food Destruction

By Andrea Hutchison

Homeostasis is defined as the state of balance among the body's systems needed to survive and function correctly. When that balance is interrupted, chaos within the body may occur and illness may develop.

Just as God designed our miraculous bodies to strive for balance American citizens strive for a similar balance called "freedom." Many who lack freedom have risked lives and left their homes and families to come to America in hopes of gaining it. Individuals who understand how the human body functions make choices to help, not hinder, the process of homeostasis, by feeding and nourishing it correctly. Those who understand American history and the suffering endured to create and keep our country free will stand up for individual rights and God-given freedom.

We Americans possess an innate desire to have and keep our freedoms, understanding our country was built upon our U.S. Constitution. It is tied to our rights to worship, educate ourselves, and keep and use firearms. Many are becoming acutely aware that something is wrong and are beginning to take a stand, realizing freedom is being stripped from our "taken-for-granted lives."

But the freedom to produce food probably crosses few citizens' minds, "It is on the shelf, isn't that all that matters?" The origin of food and the production process it goes through to arrive at the grocer's shelves is not a thought. This once highly efficient process is being high-jacked by a global control agenda targeted at removing the freedom farmers and ranchers once held to operate their businesses efficiently replacing that freedom with control. A hierarchy of "sustainable development" has been systematically injected creating a situation in which producers are being forced to comply if they wish to market their products. Compliance brings with it third-party audits and unattainable profit stripping, financially debilitating requirements. This hierarchy is replacing God's plan of individual freedom with a top-down control system and very few are paying attention.

Incrementalism, a frog in the pot scenario, fueled by public, and private partnerships between radical environmental groups, federal agencies, land grant universities, and once-trusted agriculture organizations, has made it possible for rules and regulations to unload upon our food and resource providers.

Costly social and environmental compliance hoops to prove sustainability are now required to market products proving detrimental to the survival of a rapidly growing number of independent farms and ranches. This is happening around the globe, not just in the U.S. as we have witnessed with the recent Dutch farmers boycotting this radical agenda. As producers disappear choices on grocery store shelves follow bringing higher prices and limited choices, in many cases no choice.

Whether you are a vegan or love a great steak, we should all be concerned about the freedom being stripped from our food and resource providers in the name of sustainable development. Webster's 1892 definition of sustainable (sustainable development was yet to be defined) was simple: "that may be sustained or maintained". In 1987 Norwegian "mother of sustainable development", Dr. Gro Harlem Bruntland, a devout socialist and director general of the World Health Organization along with a host of global elites re-defined this once simple term, and sustainable development became an agenda.

The Bruntland Report aka "Our Common Future" was released and sustainable development became the Trojan Horse a host for a slow-moving subtle attack to be unleashed upon our food and resource providers. Sustainable development carried with it a multitude of environmental and social regulations within air, land, water, and animals- all resources needed for food production. From regulating air quality and water usage and consumption for livestock to enforcing third-party audits to prove sustainable practices in the raising of our livestock to expanding the endangered species listing on private property from a handful of actually threatened animals to now over 5,000 listings including plants and insects, the regulations are mind-boggling and unattainable. The goal of sustainable development originators was well designed to be unattainable. The average independent food/cattle producer simply cannot survive under these regulations as they limit production on land, and increase operating expenses placing heavy financial burdens on already stressed farm operators. Besides regulations livestock producers face a balancing act of juggling extreme weather, inflationary input costs, and increased interest rates.

In 1992 during the Earth Summit in Rio De Janeiro, George H. W. Bush further advanced this destructive regulatory agenda by signing the U.S. to follow the United Nation's sustainable development plan for the 21st century also known as Agenda 21. The 40-chapter playbook lays out the plan to inventory and control all the world's resources, advancing Gro Bruntland's dark control agenda and setting the course to destroy the profitability of the world's food and resource providers.

Today sustainable development in the name of climate change and global warming vilifies carbon dioxide, a naturally occurring source required for our survival, forcing farmers, ranchers, and resource providers from energy to transportation to prove they can reduce it. Currently, elaborate social and environmental protocols must be met to prove sustainability. ESG or environmental, social governance scoring is being rolled out of financial institutes requiring a customer to fit sustainable guidelines if they are to receive access to funds. Sustainable development created Biden's 30 by 30 plan which seeks to protect 30 percent of our nation's land and water placing trust into federal hands by 2030. This plan is well on its way and now those advancing it have set a new bar- 50 by 50. The U.S. will produce nothing if this plan is left in the hands of those who are currently advancing 30 by 30.

God designed our bodies to strive for healing through a process known as homeostasis or balance, we have the freedom to make choices to nurture or neglect it, heal, or harm it. The Lord also gave us a design for providing sustenance giving dominion of land and animals over to farmers and ranchers. The freedom to choose what we feed and nurture our bodies is being replaced with an agenda of control by man, not God. Chaos has resulted. To protect our God-given freedom, education is key, we must educate others about the threats our food producers and the delicate system they utilize to get food from their farms to your plate face. The destructive force of sustainable development is aimed at humanity, it creates "unsustainability" and the battle to stop it belongs to all of us.

For more information on how sustainable development developed and what you can do to prevent further food and resource providers' destruction, visit the sites below:
- *https://www.britannica.com/topic/Brundtland-Report*
- *American Policy Center (at the time of this writing APC's website was down, but keep trying as it is one of the most thorough fact-based sites available)*
- *https://en.wikipedia.org/wiki/Agenda_21*
- *https://thenewamerican.com/print/no-farmers-no-food/*
- *https://thenewamerican.com/dutch-farmers-harness-action-to-win-big/*
- *https://americanstewards.us/issues/what-is-30-x-30/*

5 Reasons Blueberries are Superfoods

By Deborah Hanyon, MPH, RDN, ACE-CHC

More species of blueberries exist in North America than in any other continent in the world.

Blueberries have the highest antioxidant capacity of any of the popular fruits and vegetables. Antioxidants are just what they say they are: "anti", meaning they work against, "oxidation." Oxidation is a process that results in the breakdown of cells. For example, the reason bananas, apples, and potatoes become brown when open to the air is because of oxidation. Thus, antioxidants prevent cell damage in the body, an example of which is cancer. Blueberries are an excellent way to reduce your risk of cancer and other age-related diseases.

- Blueberries are low in calories. One cup only contains about 80 calories.
- These tasty fruits are a good source of fiber, providing 4 grams per serving.
- They are also high in vitamin C and are a good source of potassium.
- As with all fruit, blueberries are high in water (85%). Thus, blueberries hydrate your body naturally.
- Blueberries are rich in the phytonutrients known as anthocyanins.

Blueberries taste delicious and make many other foods tastier. You can freeze blueberries to increase their shelf life. This is great in the summer because the coldness of frozen berries makes the smoothie more refreshing on a hot summer day. Add them to your favorite cereal or to plain yogurt to spruce them up AND make them more nutritious.

Deborah Hanyon has been a Registered Dietitian Nutritionist since 2000. She earned her BS Degree in Foods and Nutrition in 1996, and her MPH in Public Health Nutrition in 1999. Deborah is also a certified Health Coach and Group Fitness Instructor with the American Council of Exercise. She also attended the Institute in Creation Research. At HomeschoolingDietitianMom.com, Deborah provides nutrition education materials for all ages, as well as for homeschooling. With a husband who is a professional cook and a son who is a picky eater, Deborah's passion is to others learn how to be all they can be health-wise, no matter what the scale reads, and to help you teach your kids how to make better health choices, no matter what their age or unique needs. You can also find her on Instagram @debbiehanyon and she is the Homeschooling Dietitian Mom on Facebook.

What Does it Mean to Know Thyself ?

By Christina Thompson

What does it mean to know thyself? I sit outside contemplating this. Around me, I hear the sounds of laughter and my grandchildren playing. The wind rustles through the unfallen winter leaves, and birds chirp nearby. I can even hear cars in the distance as I take the time to listen to everything around me.

There is so much that goes into knowing thyself. I feel it would take more than a lifetime to explain it. So how can I explain it with half my lifetime passed by? I can only try and see where God leads me.

I am three-in-one, a mind, body, and soul. We can see the body and we can even see part of the mind, but we cannot see the soul with the naked eye. It is what we experience. The experiences we go through good and bad are definitely part of what makes us know ourselves. We can let the negative experiences make us bitter in this life, or we can let them make us a better individual.

When we feel less than perfect, which is nearly every day for me, we can go to a peaceful place. I have found my peaceful place in singing, praying, listening to God, and even through painting a chicken coop (doing a mundane but worthwhile chore). We each have to find what gives us peace!

Woman Meditating, after 1868. Copy after Jean Baptiste Camille Corot (French, 1796–1875). Oil on fabric

Prayer and meditation are a concept everyone can partake in. You must find the peaceful places that only you can find for yourself. However, with this article, I will write a short piece about meditation and knowing thyself.

Why do meditation and prayer matter to knowing thyself?

In today's world, our whole being, body, mind, and soul, can be pulled in different directions. When you take time to reflect and pray, you're pushing all your other obligations to the side for a moment. You get to 'be' for a moment, letting go of everything.

It helps your mind with the constant worries and those go-go-go and do-do-do attitudes. It gives your soul time to just 'be'. It lets your body listen to everything around it. Plus, it helps you get in touch with and listen to what's going on with your body, your mind, and your soul. You have to stop being busy, even for a moment.

This happens when we sleep, but we also need it in our waking life. Why do you think that is? I personally believe reflection and prayer help us to focus not on ourselves per se, but on God, the world around us, and others. This is all part of knowing ourselves when we can look at our whole picture. When we meditate, we can practice doing this.

We are more. If all we do is go-go-go, we don't take the time to think of others. When we don't reflect on where we are or what we want to become then we can become stagnant. We need to meditate.

Adding prayer and reflection about others can help us to remember that we are not doing this alone. It helps us remember that others around us love us and want the best for us. We can become better people for our communities if we take out the time to pray and meditate.

So to conclude, I believe we can truly experience peace, objective reflection, and the betterment

of others through prayer and meditation. In return, we will have a clearer head when we go out to accomplish our goals. We will stop thinking only of ourselves, so others will benefit as well. I believe meditation and prayer are an integral part of beginning to truly "know thyself."

Christina Thompson is a grandma, wife, chicken tender, gardener, and veteran homeschooler. She loves singing and psychology. You can find her on Smule as ChististheWay.

Recipes

Basic Casserole

- 1 large can condensed cream soup (or your choice of white sauce)
- 2 pounds of cooked protein (meat, beans, or an assortment)
- 2 cups vegetables - fresh, frozen, or canned (drain the liquid)
- 1 pound cheese (optional)
- Topping - tater tots, mashed potatoes, cornbread batter, or biscuit dough

Preheat oven to 400 F. Pour cream soup or white sauce into a 9x13 baking dish. Sprinkle with protein and vegetables. Slice or grate cheese on top. Pour or arrange topping to cover all other ingredients. Bake for 30 minutes or until the topping is done and toasty.

Easy Baked Beans

- 1 large can of pork and beans (or vegan beans)
- 1 small onion
- 4 Tablespoons brown sugar
- 1 dash liquid smoke (or Chipotle Powder)
- ¼ teaspoon dry mustard
- ¼ teaspoon maple flavoring
- 2 slices bacon on top (optional)

Mix all ingredients together and bake in a casserole dish for 1-2 hours in a 350-degree F. oven.

Taco Soup

- 1 pound browned ground round chuck or turkey
- 1 large onion chopped
- 1 package mild taco seasoning mix
- 1 28-ounce can of diced tomatoes
- 1 8-ounce can of tomato sauce
- 1 can mild Rotel tomatoes
- 1 can of pinto beans
- 1 can of black beans
- 1 16-ounce can of corn
- 4 cups water to desired consistency

Simmer ingredients for 30 minutes.

No-Fail Pie Crust

- 1 ⅓ cup flour
- ⅓ cup oil of your choice
- 3 Tablespoons milk of choice
- 1 pinch of salt

Mix oil and milk. In a separate bowl, mix flour and salt. Add oil/milk mixture to flour mixture and stir until moistened. Do not over-mix. Roll out between sheets of wax paper. Place in a pie pan and back at 350 degrees F. for 10-15 minutes alone. Or, base the temperature and time on filling, covering the edges of the crust with foil to keep it from browning too much.

Things to preserve in Spring

- **Freeze** rhubarb, asparagus, strawberries, spinach, peas, and broccoli.
- **Can** sauerkraut, strawberry jam, and strawberry-rhubarb pie filling.
- **Dehydrate** kale and strawberries.

Just the Facts - Politics from the horses' mouth…

Senators sent a letter to the US Department of Energy (DOE) Secretary on the Biden Administration's proposed ban on the sale of nearly all gas stoves in America.

"We firmly believe in the free market and the ability of American citizens to make their own choices," the Senators wrote in a letter to the US Department of Energy (DOE) Secretary on April 20, 2023, regarding the Biden Administration's proposed ban on the sale of nearly all gas stoves in America.

"By effectively banning gas stovetops through the imposition of excessively stringent efficiency standards, the Department of Energy is depriving Americans of the freedom to choose the type of appliance that best suits their needs, whether it is for cooking, heating, or any other purpose," the letter stated. "This overreach not only infringes upon the rights of our citizens but also risks creating a potentially uncompetitive market by limiting the options available to consumers."

Senators James Lankford (R-OK), Roger Marshall, M.D. (R-KS), and Steve Daines (R-MT) were joined on the letter by Senators John Barrasso, M.D. (R-WY), Roger Wicker (R-MS), Todd Young (R-IN), Joni Ernst (R-IA), Jim Risch (R-ID), Cindy Hyde-Smith (R-MS), Markwayne Mullin (R-OK), John Hoeven (R-ND), Ted Cruz (R-TX), and Bill Cassidy (R-LA).

On April 19th, 2023, Senator Lankford questioned Internal Revenue Service (IRS) Commissioner Daniel Werfel in a Senate Finance Committee hearing about the "historical level" the IRS plans regarding audits, particularly regarding potential audits for those making less than $400,000.

"Republicans actually brought an amendment saying it can't include that $400,000," Lankford said. "Democrats actually voted that amendment down and blocked it."

Lankford also voted on the joint resolution of disapproval under the Congressional Review Act (CRA) to nullify an illegal US Department of Veterans Affairs (VA) rule that provides abortions through the taxpayer-funded VA health care system, even in states where unborn life is protected.

"Service members are heroes and should get the care they need and deserve from the VA," Lankford said. "It's appalling that the Biden Administration has diverted VA healthcare funding from veteran care to abortion."

The CRA gives Congress the authority to review major rules issued by federal agencies. When the rule was issued last year, Lankford led 67 colleagues in filing a public comment letter to call out the VA for trying to go around federal and state laws and provide abortions. The resolution failed by a vote of 48-51.

"Abortion is not health care, and taxpayers should not be funding it, especially at our VA health centers. Long-standing federal law is clear, the VA is not allowed to provide abortions.," Lankford said. "The Senate should speak out when the Administration ignores the law because of their obsession with abortion on demand."

Lankford also issued a statement Supreme Court's action in continuing to allow chemical abortions.

"I'm disappointed that the Supreme Court has allowed chemical abortions to continue, even with the FDA ignoring the clear statute," Lankford stated. "I will continue to do everything in my ability to protect women, girls and unborn children from the real risks of DIY-chemical abortion pills."

On April 19th, Congressman Frank Lucas (OK-03), Chairman of the House Science, Space, and Technology Committee, joined colleagues on the House Agriculture Committee for a hearing with U.S. Environmental Protection Agency (EPA) about choosing to ignore evidence regarding chlorphyrifos use.

"I have seen issues arise when Agency scientists at USDA and EPA are at odds," Lucas stated. I respect differences of opinion. But when the science is generated by people closer to the issue in the field and the use, I think we have to give them the benefit of the doubt."

Three Case Studies & What We Can Learn From Them

By Myrna Stiles Buckles

I will never forget one of the most traumatic periods in my life! The school year had just started for the fall and my parents moved us to another town 150 miles away! It was just two weeks into the school year!! I was in a panic because I was a shy child and had never considered moving.

To add to the sick stomach I already had from abruptly moving, we moved into my grandparents' former house which had been unoccupied for about 10 years. This house had been built in the late 1800s for a Choctaw Principle Chief and was bought in 1922 by my great-grandfather's two-story.

It was a two story house, all wood floors that cracked and popped when you barely took a step. Not only that, there were bullet holes in at least 2 places from the lower level up to the floor of the upper level. This was just about too much for this shy little girl. It was a great relief when my parents set up the bedroom just above theirs for my bedroom. The stairs from that room went down directly into their bedroom.

The first night we were all in bed with the house quiet when all of a sudden we heard what sounded like an army running up and down the stairs that led from my room down to my parents' room. My eyes flew open and then we all heard my mother demand, "LEW-IIIIISSSS!" It was so frightening to hear those huge rats running up and down the stairs and my mom wanted them out of there that night.

I am so thankful for my mom. She made sure we had what we needed and that we knew people around us as soon as possible after that move. She was amazing at locating the community that was already there and creating a community that we needed.

The Bible tells us in *Romans 15:4* *"For whatever things were written before were written for our learning, that we through the patience and comfort of the Scriptures might have hope."*

God has gifted me with a variety of experiences fostering community among a group of people. I will be sharing some of them while looking at their similarities, differences, and what I have learned from each of them.

My generation grew up without cell phones, the internet, computers, social media, etc. This gives me a perspective that allows for some observations of the differences when fostering community when we were more likely to be communicating face-to-face versus electronically. Through this series of articles, I will share experiences/case studies of each and tips for you to use in your day-to-day life.

Case Study 1: Tiner Community

When I was in the 6th grade, my family moved from a town, Holdenville, where most everything was within walking distance, 150 miles away to my grandparent's home place that was 6 miles from the nearest town of any size. We quickly learned that there was a community group in the surrounding area. I have learned over the past 50+ years just how important that group was for my family and others.

The community group was called Tiner Community. Tiner was the name of the nearby one-room school. The community maintained the historical school building and used it for community gatherings.

That one-room school was a big part of what brought people together. They took turns mowing the grounds, maintaining the building and gathering there.

Looking back, I find it interesting that a building could do that but it was more than a building. It was a cause, a universal cause of the people living there that brought them together. Preserving the building and its history was the cause and they all agreed on it.

They held monthly birthday parties, game nights, swapped garden produce and preserved foods. They also helped one another move cattle, mow yards, and provided meals for one another when someone was ill.

Our closest neighbors to the south had land on both sides of a fairly large river in SE Oklahoma and a herd of cows of about 40 head. The only way to move the cows from one pasture to another was across a busy highway bridge where the speed limit was 70 mph until highways across the nation were changed to 55 mph.

When they were ready to move cows, they typically called us and my brother and I would go help them. We often were on foot pushing those cows across the bridge and sometimes one of us was on a horse. I was always a little anxious about getting that project done but we did quite a few times over the years.

I also mowed the grass in their yard. I loved it because they treated me like I was important (I was the youngest kid in my family.). I remember getting the best glass of iced tea and some chocolate chip cookies during the process. It was a win win.

In later years, after my mom was no longer living, the neighbor ladies made my dad his favorite jelly from wild muscadines and he always made sure to have a quart sized bag of shelled pecans ready to give to them. I remember getting the best glass of iced tea and some chocolate chip cookies during the process. It was a win-win.

In later years, after my mom was no longer living, the neighbor ladies made my dad his favorite jelly from wild muscadines and he always made sure to have a quart sized bag of shelled pecans ready to give to them.

Case Study Comparisons

Attribute	Case Study			Conclusions
	Tiner	CRIT	RBP	
Cause or Rallying point	X			
Exchange of Perceived Value (work, food, etc.)	X			
Purposeful Intention	X			
Consistent Action	X			
Commitment by Those Involved	X			
Opportunity for Interaction	X			

I lived this scenario and over time began looking at it from a higher observation point after my mom died. She was diagnosed with ovarian cancer in September and died the following June. Neighbors and friends brought meals to them all the while she was sick and many of them up to two years afterward. That told me I needed to pay attention to what they did to build relationships and community.

Case Study 2: Injury Prevention Coalition - Colorado River Indian Tribes (CRIT)

As a Public Health Service Officer, I was assigned to the Indian Health Service on the Colorado River Indian Tribes Reservation from 1998-2004 as an Environmental Health Officer. This is a position that serves the tribes to prevent injuries and illnesses.

Injuries are a leading cause of death among Native Americans and we were tasked with reducing injuries. This was an overwhelming task and I knew I needed community involvement. After reviewing those I had a relationship with in the community, I set an intention to contact those I thought would be important leaders that could impact injuries in their community. People like the Tribal Health Director, Tribal Police Chief, Head Start Director, and others.

I began making it a practice to mention the idea of an Injury Prevention Coalition each time I had a meeting with those I selected over a period of 3-6 months. Once they were all on board, I set up a meeting where I got input from them about starting a coalition.
A Community Coalition establishes and builds relationships within a community where better ideas and outcomes can be achieved together. (Community Coalition - SCDHEC)

Attribute	Case Study			Conclusions
	Tiner	CRIT	RBP	
Cause or Rallying point	X	X		
Exchange of Perceived Value (work, food, etc.)	X	X		
Purposeful Intention	X	X		
Consistent Action	X	X		
Commitment by Those Involved	X	X		
Opportunity for Interaction	X	X		

After that initial meeting we as a group organized and started making a bigger impact in the community. We were able to get a smoke detector program for head start families, obtain a grant to support community injury prevention activities and developed a solid program empowering local leaders to continue into the future.

Case Study 3: Ring Bomb Party Business & Community (RBP)

In October of 2019, I started a business through a direct sales company (Ring Bomb Party). The business was dependent on facebook live sales and building an audience of loyal fans/buyers. These tasks were quite challenging in the beginning and became a lot of fun in the end.

I personally got a tremendous enjoyment from the lives because I got to dress up in beautiful sequin tops and throw a party without cleaning my house or cooking food and my husband didn't have to endure a party at our house. He isn't a party kind of guy. It served us well as a couple.

I employed many of the skills I had learned from my parents and from me working in Native American communities to build the audience aka community. We played games on the lives that built camaraderie and fun; I paid close attention to the comments on the live and interacted with them

Attribute	Case Study			Conclusions
	Tiner	CRIT	RBP	
Cause or Rallying point	X	X	X	Each case study included each attribute though at different levels and they were often applied differently.
Exchange of Perceived Value (work, food, etc.)	X	X	X	
Purposeful Intention	X	X	X	
Consistent Action	X	X	X	
Commitment by Those Involved	X	X	X	
Opportunity for Interaction	X	X	X	

Conclusion

In each case study, the following attributes were present though in different degrees and with different applications:

- Common Cause or Rallying point
- Exchange of Perceived Value (work, food, etc.)
- Purposeful Intention
- Consistent Action
- Commitment by Those Involved
- Opportunity for Interaction

You've likely heard the saying, "there is power in numbers" and it is so true. Or, "many hands make light work". Both of these skirt around the details of bringing a community together though they do acknowledge the need. Community is so much more than either of these quotes indicate.

Community is so strong because of the way its members work together to bring their unique collection of resources together. The uniqueness of each community is dependent on the skills and talents of its members. One of the most important strengths of community is when people put their God-given talents together, the results can be exponential. God wants us to be in community with others.

> Romans 12:4-5 *"or as we have many members in one body, and all members have not the same office: So we, being many, are one body in Christ, and every one members one of another."*

It is important to note, there can be negative aspects of a community as well as positive ones. The key is to develop the community with structure and leadership. Each situation determines how much structure and leadership is needed.

The Adventure of Homeschooling

Summer 2023   PrairiDustTrail.com   Volume 01 Issue 03

PrairieDustTrail.com

Connect with the past ~ Prepare for the future ~ Live more sustainably

The Adventure of Homeschooling

- Tips & Tricks from veterans
- Financinces and Homeschooling
- Advice from homeschool graduates

Editor's Epistle

By Dawnita Fogleman

Homeschooling has been one of the most rewarding adventures of my life. I really don't understand career women. At one point in my life, before I met Paul, all I wanted was to be self-sufficient. Having a career was never my goal. I figured it was a means to an end, nothing more.

As a survivor of abuse, I have always been highly protective of our children. The very thought of sending them to an institution for eight hours a day where I had been bullied by peers and had some pretty hard and uncomfortable experiences was enough for me to start searching for options. I had never heard of homeschooling before. We first looked at the area Christian school. While lovely, the cost and distance to travel would have required me to get a full-time job, which we agreed it wasn't a good option for us.

I don't remember where or how I heard of homeschooling, but it immediately answered my prayers. I never doubted I could do it. I was determined.

Like most new homeschooling moms, I made our little classroom look like something I remembered from first grade, coupled with a romantic look at the Little House on the Prairie or Anne of Green Gables one-room-schoolhouses. I have a photo of our oldest crying that first year at her desk. I snapped that photo because it was a moment I never wanted to forget, the moment we actually began homeschooling, and quit schooling at home.

The transition took years. We were the first homeschoolers in our county and I knew all eyes were on us, judging every move. The longer we journeyed, the more relaxed it became. Within about five years, the children didn't realize how much school was part of our life. They only saw school as the reading books, handwriting journals, and the dreaded math books. Actually, what we were doing was lifelong learning, exploring the world around us and all the possibilities it held.

My only big requirement for our homeschool was for everything to be Biblically based. We started by building a model tabernacle. The reading books were Bible stories. And of course, HIS-story and Creation science.

So often, we make things harder on ourselves because of perceived expectations. YHWH's expectations are really the only ones that matter. Reading and following HIS word is most important. If we can teach our children to do that, we've succeeded.

The Manual for Children

By Kathryn White

"And, ye fathers, provoke not your children to wrath: but bring them up in the nurture and admonition of the Lord." Ephesians 6:4

When God gives us a gift, He doesn't just drop it in our lap and leave without a word. He actually teaches us what to do with it. And, in the case of children, that is no exception. He has a lot to say about it AND He leads by example.

If we've been in church at all, then we've all heard Ephesians 6:4 at least once. But, let's pause and think about it for a moment: what does it mean not to provoke your children to wrath?

Wrath means extreme anger, annoyance, or - my favorite word - exasperation.

Here are 4 ways we as parents might provoke our children to wrath:
1. Set impossible standards so that a child despairs at ever achieving them.
2. Tease, ridicule, or humiliate a child as a means of punishment.
3. Be inconsistent so a child is never sure about the consequences of their actions.
4. Be a hypocrite and require behavior from your child that you are not daily setting an example of yourself.

Now, let's move on to the second part: "but bring them up in the nurture and admonition of the Lord". This simply means that parents should train their children the way God trains us. God sets a perfect example for us all through scripture.

As a Father, God does 4 things consistently:

- He is slow to anger (Numbers 14:18)
- He is patient (Psalm 86:15)
- He is forgiving (Daniel 9:9)
- He gives appropriate discipline to draw us to repentance (Hebrews 12:6–11).

The word discipline comes from the root word disciple. To discipline someone means to make a disciple of him. God's discipline is designed to "conform us to the image of Christ" (Romans 8:29). As we are consistent in our daily practice of a godly lifestyle (Galatians 5:16, 25), we can encourage our children to follow our example, just as we follow Christ's example. (Ephesians 5:1)

The ultimate goal of "bringing up" our children is to produce wise people who know and honor God with their lives.

> "The father of the righteous shall greatly rejoice: and he that begetteth a wise child shall have joy of him." Proverbs 23:24

As the highs and lows of the coming school year take their toll, remember to stay in the scripture daily with your children, and remember that you too are a child - a child of God - His disciple.

Snake Identification and Care for Bites

Compiled by Prairie Ruth Fogleman

Venomous snakes in the United States consist of rattlesnakes, copperheads, cottonmouths, and coral snakes, according to CritterControl.com. These belong to a group called pit vipers, named for deep depressions found on either side of their heads halfway between their eyes and nostrils.

A venomous snake's scales tend to appear in a single row on the undersides of their bodies and tend to be more pointed or of a triangle shape. Harmless snake species have two rows of scales.

Water snakes can be harder to identify because they also tend to have diamond-shaped heads, which is typical of venomous species, but they aren't venomous.

(Photo: Mojave Rattlesnake (Crotalus scutulatus). Cochise County, Arizona. Spring 2023.)

Pit vipers have vertically elliptical, or egg-shaped, pupils that may appear slit-like depending on the lighting, while most non-dangerous species of snakes have perfectly round pupils.

When heading out into the wilderness, it's a good idea to take precautions for the possibility of running into a snake. Most are harmless and uninterested unless surprised or provoked.

(Photo: Eastern Copperhead (Agkistrodon contortrix). Anderson County, Texas.)

Great-great Granny always carried a big walking stick with her in case of running into a snake. It's also a good idea to wear boots or high-ankle hiking shoes. Be aware of your feet as you are traveling and avoid thick grass or brush whenever possible.

If you do happen to get bit, here are some basic first-aid tips from the USDA:

Snakes on the Plains Photography

DO NOT:

- Do not make incisions over the bite wound.
- Do not restrict blood flow by applying a tourniquet.
- Do not ice the wound.
- Do not suck the poison out with your mouth.
- These methods can very well cause additional harm and most amputations or other serious results of a rattlesnake bite are a result of icing or applying a tourniquet.

(Photo: Northern cottonmouth, Agkistrodon Piscivorus.)

DO

- Stay calm
- Call Dispatch via radio or 911
- Wash the bite area gently with soap and water if available
- Remove watches, rings, etc., which may constrict swelling
- Immobilize the affected area
- Keep the bite below the heart if possible
- Transport safely to the nearest medical facility immediately.

Photos by "Snakes on the Plains" from Duncan, Oklahoma. You can find more on Youtube @SnakesonthePlains, Facebook Photography - facebook.com/snakesontheplainsphotography Instagram @snakesontheplains or email snakesontheplains@gmail.com for more information.

Homeschooling in a Shoe

By Israel Wayne

"There was an old woman who lived in a shoe.
She had so many children, she didn't know what to do."

This line from an old nursery rhyme, originally written in 1794, evokes a mental image of a woman at her wit's end. She is so frazzled and overwhelmed, that she may very well respond badly to those within her care (as the rest of the story goes on to indicate).

The sad thing is that not much has changed in 200+ years. If anything, it seems that life today is far more complicated. Technology, which is supposed to simplify our lives and make things easier, seems to instead offer us a myriad of ways to be distracted from truly important things, and to create a continuous and monotonous cycle of repair and maintenance.

Stressed-Out Parenting

Most of us live with very little margin in our lives. Even as homeschoolers, we often book our every spare minute full of activities. Today, more than ever before, there are endless social options for our students. Co-op classes, field trips, church activities, civic events, advanced study lessons, seminars, conferences, camps, retreats, etc., are all being extended to us, as a way to supplement and improve our family relationships and the advancement of our child's education.

None of these options are bad, in and of themselves. The problem is in the excess. Sometimes, we just don't know how to say no, and we over-commit.

Fear-Based Parenting

Much of the frenzy, in my view, is based in a fear of failing our children. We believe that our children may miss some important skill or life lesson, and they may therefore grow up educationally malnourished and it will be all our fault. We have scary visions of our children sleeping under the overpass, and spending their days on a street corner with a sign that says, "Will work for food!" All of this, because we neglected to enroll them in that special Constitutional Law class being offered by the local co-op. So, we run here and there, endlessly trying to cram it all in before our children graduate and the door of all future learning slams shut on them.

Homeschooling the Herd

My wife and I currently have eleven children. That is considered to be a lot. At least that is what I've surmised by the expressions I see on the faces of those we meet at the grocery store. "Are these all yours?!"

"No, I only have five with me today. The other six are at home." This is met with predictable responses such as choking, clearing the throat, raised eyebrows, low whistles, and comments such as, "Have you figured out what causes it?" or "Are you trying to start a baseball team?"

I never want to tip my hand and reveal that really we are just in the beginning stages of world domination, which will begin with us buying our own remote island somewhere, where we will exert taxes and duties on any unsuspecting victim who happens to be shipwrecked and wash up on driftwood.

If you study the history of ancient civilizations, you will discover that every successful dynasty, whether the Sumerians, or Egyptians, or Persians, or Romans, or Incas, or Mayans...all of them began with some guy and his wife who had eleven children. This is a little-known fact that has been suppressed for some reason by all major textbook publishers. My heart, however, tells me it is true.

But for now, while we await our future global exploits, that will be chronicled by historians yet to come, we are just trying to keep our house from falling apart and stay on some kind of a decent academic schedule. Perhaps you find yourself in the same boat.

Whether you are homeschooling one or two children, or an entire baseball team full of future world conquerors (as we are!), you are battling stress and, in many cases, anger, as you try to keep your head above water.

Tips for Survival

Here are some homeschooling survival tips that we have found to be useful when you find yourself drowning in life and homeschooling.

1. Have a regular date night.

Okay. We aren't actually very good at this. But we try. For us, going out regularly to dinner isn't very practical, so what we've taken to is evening walks around the neighborhood.

As parents, we need exercise and fresh air and it gives us a chance to reconnect and talk about what we are doing well, and what we are doing poorly. If you are a single parent, schedule time to meet up with a homeschooling friend and compare notes. You need to constantly reassess your progress and make sure you are on course with your goals.

2. When possible, teach multiple grades at once.

If you are teaching the older ones about the Revolutionary War, find a way to incorporate videos, books that you read aloud as a family, and field trips for the younger ones as well. It is true that the little ones won't remember much, but they are being introduced to the topic, and it will lay a foundation for them that will be helpful when you cycle back around to it in their later studies. The main thing is that everyone is learning something and you are doing life together.

3. Never lose sight of the big picture.

Never forget WHY you are homeschooling. It is because you want your children to know, love and serve God, and love and serve others. All of the teaching and instructing you are doing is for that end purpose.

4. Declutter.

We are definitely not good at this! But we are getting better. We have found that massive inefficiency happens in our lives because we just own too much stuff for our living space. We spend more time hunting for things than we should, we own things that don't get used, and things are even more likely to be broken, all because we just have too much stuff. We are in a massive downscaling effort right now. Sell it, give it, donate it, repurpose it...but don't just store it. If you don't use it, get rid of it! It is essential to sanity.

5. Create a schedule and stick to it.

Once again, this is a challenge for us. But we find that we all do better (especially our children) when they have some structure and routine to their day. When they are drifting, not knowing what to do, they find their own ways to entertain themselves, and that always goes bad! Even playtime needs to have some structure and time parameters. At least, this is what we have found works best for our family. Getting enough rest is also critically important. Even if "early to bed, early to rise," doesn't make you "healthy, wealth and wise," as Benjamin Franklin predicted, it will generally lead to better focus and happier attitudes throughout the day.

6. Don't be a slave to your curriculum.

Part of the wisdom portion of parenting is knowing when your curriculum is serving you well, and when it is a tyrant. If your curriculum isn't working for you, don't be afraid to sell it and start over. Get the tools that are right for your family. Maybe what you bought works great for your friend at the co-op, or was applauded in the online review you read. But your child is unique, and your family isn't like anyone else. Be willing to think outside the box and do what you believe will work best for your child.

7. Nurture relationships.

Don't forget to have fun in the process of parenting. If you have made parenting all about work, study, chores, and rules, you will drive your children away from you. You need to make sure that you are scheduling plenty of family enjoyment time as well. Have fun just for the sake of having fun. When schooling creates tension in family relationships, work on the relationships, and then cycle back to the schooling.

Remember, you aren't the first person on the planet to walk this road. Thousands of other parents have gone before you and have successfully homeschooled their children, even on meager incomes, crowded spaces, limited resources, and health struggles. It can be done! You just need to have clearly outlined goals, a willingness to make changes as needed, and a tenacious commitment to stay the course and do what is best for your child.

Instead of being a tragic story, the woman in the shoe can be a picture of great contentment and happiness, fulfilling an eternal calling.

> *"Thy wife shall be as a fruitful vine by the sides of thine house: thy children like olive plants round about thy table." (Ps. 128:3)*

> *"The Lord is the portion of mine inheritance and of my cup: thou maintainest my lot. The lines are fallen unto me in pleasant places; yea, I have a goodly heritage" (Ps. 16:5-6).*

Israel Wayne (www.IsraelWayne.com) is an author and conference speaker. He is a co-founder of Family Renewal (www.FamilyRenewal.org) and is Site Editor for www.ChristianWorldview.net. He is the author of the books: Education: Does God Have an Opinion?, Raising Them Up: Parenting for Christians, Pitchin' a Fit: Overcoming Angry and Stressed-Out Parenting and Answers for Homeschooling: Top 25 Questions Critics Ask (among other titles). He and his wife Brook are both homeschooled graduates and have homeschooled all eleven of their children from birth in southwest Michigan.

What is Turmeric Good For?

By Deborah Hanyon

I've been hearing a lot about turmeric and curcumin. However, as a Registered Dietitian, I hold to a professional code of ethics standard. This requires that I be cautious when recommending supplements used for medicinal purposes. As a result, I ensured my own knowledge was sound before sharing it with my readers. This article will discuss, "What is Turmeric Good For?" and, "Whether it is safe for everyone."

Turmeric has antioxidant properties. This is related to the fact that it is a plant.

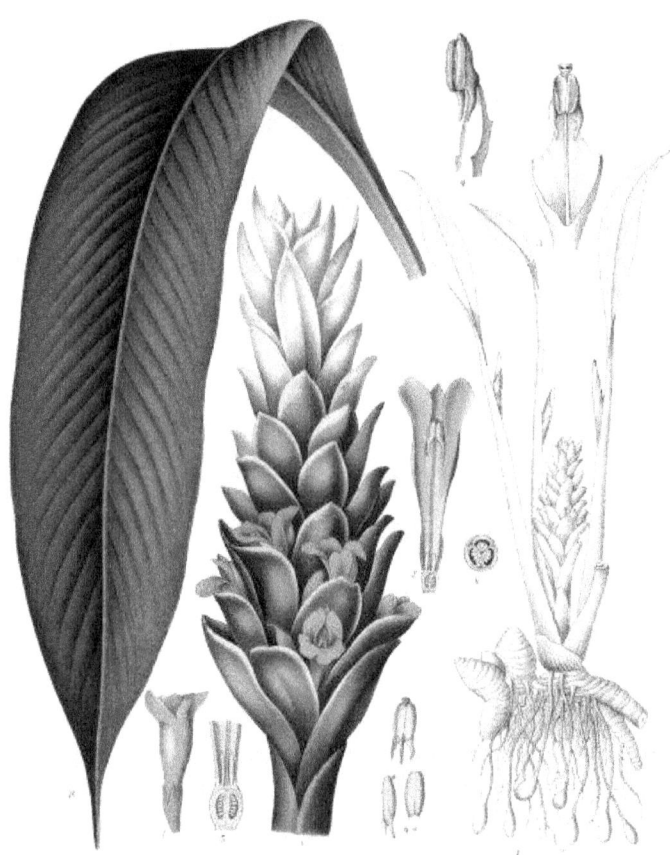

What is Turmeric?

Turmeric is a type of root plant related to ginger. It is found in curry seasoning and is commonly used as a spice in Indian cuisine. In Asian dishes, it imparts a mustard-like, earthy aroma and pungent, slightly bitter flavor to foods.

Strictly speaking, the color comes from curcumin, the active ingredient in turmeric. This root is used to dye certain fabrics in India as well as for food coloring. Examples of some foods colored using the dye from curcumin are cheese, salad dressing, butter, pickles, and mustard. Curcumin is also responsible for the therapeutic benefits of turmeric.

(Public Domain Image: Curcuma longa, Franz Eugen Köhler, Köhler's Medizinal-Pflanzen)

Why is Turmeric Good for You?

According to WebMD, turmeric is good for hay fever, depression, high cholesterol, osteoarthritis, and itching (pruritus).

- Hay Fever - Curcumin appears to reduce the symptoms of hay fever, including sneezing, runny nose, and congestion.
- Depression - Curcumin has been shown to reduce symptoms of depression in persons already taking anti-depressants.
- High Cholesterol - Triglycerides, a type of blood fat linked to heart disease seem to be positively affected.
- Osteoarthritis - Some research shows that taking turmeric extracts may reduce pain in those with knee osteoarthritis.
- Itching (Pruritus) - When taken as directed, turmeric is effective in easing itching in persons with kidney disease. Also, black pepper increases the absorption of curcumin powder. Thus, combining curcumin with black pepper helps chronic itching in persons exposed to mustard gas.

Is Turmeric Safe?

It should be noted that not every plant is safe for human consumption. However, turmeric is likely safe when taken as part of regular food, as a food seasoning, for example.

The root is used in medicine for numerous treatments. As a result, people will sometimes think, "If a little is good, a lot is better." This is most definitely NOT the case.

For example, WebMD states that turmeric is safe during pregnancy when ingested as part of regular food. However, it is likely NOT SAFE to ingest medicinal amounts during pregnancy.

It is vitally essential that dosage instructions and drug-nutrient interactions be considered when deciding whether to take or how much to take.

Drug Interactions

Turmeric slows blood clotting (increased blood thinning). As a result, it interacts with other blood-clotting medications including aspirin, ibuprofen, Plavix, warfarin, and heparin. So, it is important to check with your healthcare provider if you are taking any prescription medications to ensure they are safe to use.

Financing Homeschool Frugally

By Danielle Tate

"So, how much does it cost to homeschool?" An acquaintance asked me that not too long ago. What a loaded question!

It costs a great deal of my time... hours of reading, watching, planning, exploring curriculums, and bettering myself to be the best homeschool mama I can be.

But that wasn't what she was asking....she wanted to know how much money I spend on homeschooling because the cost is important (well, not really, but more on that later).

And if you're a fellow homeschool family, I know you're thinking about next year right now. What do you have that can be reused? What consumables do you need? And paper... do you have paper for the printer? Will you bind your own books or hit up the office store and have them do it for you?

Most homeschool families I know tend to be on the frugal side. But let's face it, you will spend money homeschooling in one way, shape, or form. Even if it is extra gas running to the library to borrow books, you'll spend money.

But what should you spend on homeschooling each year? If you're getting started, how do you know if you're buying enough and when have you gone overboard?

I'm in a dozen or more homeschool groups on social media and these questions are asked over and over by veterans and new homeschoolers alike. Today, I'm not going to answer either of those questions. Because what you spend is arbitrary and very subjective based on your family's needs. The better question to ask yourself is this, "What do I need for my child to be successful and reach the goals I (or we) have set for this school year?"

Then, seek the Lord and let Him guide you.

Each year in the Spring I make a tentative plan of what I might need/want based on subjects and interests. Then I sit down with God and work through what He is highlighting for us. I also ask my son about his interests and what he would like to do. This year he hit me out of left field with learning pottery for his art class. I never would have guessed that, but he's interested so we'll go with it.

As much as I want to jump in and buy everything on my list, I have learned (the hard way) to wait. And I encourage you to wait too. The biggest way you can save on your homeschool expense is to resist the urge to buy it all right away.

Mull over your choices for a minute. Let them sit on paper or in the spreadsheet. Not to decide if they are the unicorn curriculum you've been dreaming of (you're not going to find that), but to be sure that now is the time you need it and that it will suit the needs of your child(ren). Don't let FOMO (fear of missing out) make you rush into purchases you don't need now.

Another simple thing you can start doing right now is setting money aside for your school needs. Each month, all year long, tuck away a "homeschool savings" sum so that you have money throughout the year. You do not HAVE to buy every book you need in August to start school in September. Set money aside each month to use when you need it. Keep a tab on what your expenses are throughout the year so you can adjust your money amount accordingly.

If money really feels like a stumbling block to getting started, let me encourage you to seek and use free resources. Don't let money hold you back or fear of overspending. You will buy stuff you don't use, you will buy things that don't work for your family, and that's okay. You're not a failure. That's what used curriculum stores/sales are for.

Here are a few other practical ideas for keeping your homeschool expenses low:

- Use the library
- Use documentaries (YouTube, Prime TV, etc, etc)
- Seek out free curriculum (like Ambleside Online)
- Don't print every blessed paper you come across...use the tablet, computer, or project it to the TV
- Buy used...there is no shortage of places to buy used curriculum in person or online

- Sell your old/unused curriculum and books
- Use LibriVox fur publish domain audiobooks
- Hit up thrift stores and yard sales for equipment or resources like microscopes, globes, etc
- Hit up your homeschooling friends for things you can borrow

As you venture into the new school year, resist the urge to both buy it all and fear you'll overspend. You'll do both from time to time so give yourself grace and a monthly homeschool savings fund to help you out.

Happy homeschooling!

Danielle Tate is a personal finance coach. Her passion is to see people live free from past financial mistakes and thrive where they are on the way to reaching their financial goals. Danielle helps Christians create a finance plan that will cover monthly & unexpected expenses, reduce big-bank debt dependency, honor God, and stop money fights. Get started today with my free Six Steps to a Debt-free Lifestyle eCourse at DanielleTate.org/freedom.

Essential Tools for the Homestead

By Ryland Vassar

All homesteaders know the right tools get the job done. But what tools are essential for homesteaders? Drawing from my own experience as a homesteader and a woodworker, I would recommend the following tools as essential for your homestead.

Farm Tools:

- Shovel - You'll have to dig holes on your homestead. A spade-ended shovel will help you do just that. You can also get a flat or square-ended shovel to help you move dirt in or manure around.
- Rake - Whether raking leaves to compost into leaf mold or cleaning out a chicken coop, a rake will help you get the job done.
- Post-hole digger or T-post driver - If you are building fences, these will help.
- Wheelbarrow, utility cart, or wagon - You'll need to move something around on your homestead at some point. That's where a wheelbarrow, utility cart, or wagon comes in. If you get a wheelbarrow, I recommend the two-wheeled kind. It'll be easier to balance and push when full.
- Loppers - You'll run into an unwieldy tree at some point. Take care of its thinner branches with a pair of heavy-duty loppers.
- Axe - If that unwieldy tree just has to come down, an axe is your perfect companion. It's also great for chopping firewood and chicken heads off. Plus, it's great exercise.

Hand Tools:

- Hammer - This is one of the most common, most needed tools. You'll get tired of using rocks eventually, so might as well get one now.
- Pliers - Useful for pulling out staples, bent nails, and other wonders. There are 25 different types of pliers, but you can start out with alligator pliers for everyday work and needle-nose pliers for small spaces.
- Tape measure - Useful for measuring distances and wood. Just remember to measure twice, then cut once. You can always take more off, but it's really hard to glue it back on.
- A variety of fasteners (screws, nails, and staples), hinges, and handles - These are super important to keep on hand.
- Fencing pliers - These are handy for stretching, twisting, and cutting wire. They also have a little hammer and claw on the end for driving in nails and pulling out staples.
- Wire cutters - These are helpful in so many situations. Did your goat get caught up in the fence? Wire cutters to the rescue! Find a loose wire liable to poke your chicken's eye out in their coop? Snip it!

- Hand saw - A hand saw is important for quick woodwork out in the field. It's a great non-powered option.
- Chisels - Saws don't always make the cuts you need. A chisel can help you start cuts, notch put joints and hinges, and makes cuts in small spaces that saws can't.
- Stapler - Great for securing tarps to your chicken coop or for putting up light or temporary fencing.
- Utility knife - Great for cutting twine, opening bags of soil, compost, or feed, and cutting zip ties and rope.
- Multi-tool - My favorite and most used tool. Everyone needs one of these whether homesteading or not. Most normally consist of pliers with built-in wire cutters, Phillips and flathead screwdrivers, a knife and a saw blade, and a file. You can use it in so many ways!

Power Tools:

- Chain saw - This works better and faster than an axe for clearing trees. Just don't get carried away.
- Drill - Use it to drill holes in just about anything. Just be sure to use the right drill bit.
- Driver - Hand screwdrivers are great, but this will make your work much faster.
- Circular Saw - use this to rip sheet goods (plywood, OSB, roofing materials, wallboard, etc.) into the size needed.

Other:

Tool chest or belt - Use one or both of these to carry your tools around the homestead so you won't are less likely to misplace them.

Another tool I recommend is a portable table saw. It's not essential, but I have found it valuable in homesteading. It improves the accuracy of larger cuts when using it versus a circular saw.

You will learn the need and convenience of more specialized tools as you start with the basics and progress in your journey.

I have a method I've used for my collection of tools. If I see a tool on sale for a deep discount somewhere and my budget allows, I buy it. You can check stores like Tractor Supply or your local feed stores, garage sales, estate sales, thrift stores, or salvage stores.

Now go and build your tool chest!

Ryland Vassar is husband, father and homesteader. He is also a legally blind maker. You can follow his adventures at @r.c.makery on Instagram.

Recipes

from the kitchens of Frances Stiles via Myrna Buckles

Mom's Favorite Fruit Salad

Ingredients
- 3 Fresh Red Delicious Apples, cored and chopped into bite size pieces
- 3 Fresh Oranges, peeled and cut into bite size pieces
- 1 small can crushed pineapple
- Chopped pecans to taste
- Shredded coconut to taste

Mix all ingredients together and chill.

Frances' Blackberry Cobbler

Ingredients
- 2 qts Blackberries (fresh or frozen)
- 2 ⅔ c. Sugar
- ⅔ c. Flour
- 2 T. Butter

Mix sugar and flour well. Then place berries and sugar/flour mixture in a pan and cook berries over medium heat until thick. Stir often.

Roll out and place one pie crust in the bottom of a 9X13 inch cake pan covering the bottom of the pan and up the sides. Pour filling into the crust. Cut the butter into chunks and place on the top of the filling. Add top pie crust and cut slits in the crust to allow air to escape. Sprinkle the top of the crust with sugar.

Bake at 400 degrees F about 30 minutes until the crust is browned and the pie filling is bubbling through the slits.

Shepherd's Pie

Ingredients
- 2 pounds Ground Meat - beef, deer, etc.
- 2 cans Green Beans
- 6 large potatoes
- 1 Onion
- Butter (amount is your preference)
- Milk (amount is your preference)

Heat oven to 350o F. Chop Onion and cook it with ground hamburger. Break up ground meat while cooking. When done, spread in the bottom of a 9 x 13 cake pan. While meat and onion are cooking, put two cans of green beans in a saucepan and cook until most of the liquid is gone. Spread the green beans on top of the cooked ground meat in the cake pan. Peel and boil 6 large potatoes. When they are easily pierced with a fork, drain and add milk and butter. Then using a mixer or potato masher, mash the potatoes. Spread the mashed potatoes on top of the other ingredients in the cake pan. Bake for 15-20 min until mashed potatoes start to brown.

Optional Additions and Alternatives:
Add shredded cheddar cheese to the top and cook until melted.
One friend of mine said they put mashed potatoes in the pan first, add cooked ground meat, corn, and then mashed potatoes on top.

This is a recipe you can definitely make your own based on your family's needs, food allergies, food you have on hand, etc.

4 Reasons Every Homestead Should Have Quail

By Francis Roland, Youth Contributor

If you are just starting with farming and are looking for a simple livestock project which will give you good food that is self-sufficient, cheap, and can also make you money, keep reading. Even if you've been farming for a while I think quail are a very good idea for 4 reasons.

#1 Ease of Care

Quail are very easy to care for and raise. When you first get chicks you will see that they are very hardy. Most chicks easily survive the first few weeks. Quail grow very quickly and can be moved out of the brooder usually in 2-2.5 weeks.

Once moved to their new coop, they need very little care. With the right setup, you will only have to take care of them once per week, aside from collecting eggs.

Adult quail (4-6 weeks old) are very hardy as well. They can easily survive -20 * F - 100* F if they have protection from wind and some shade.

#2 Space

Not all of us have 100 acres. Most people live on a half acre or less. If you want to have poultry for eggs and meat, it is better to get quail than chickens. Chickens need about 6 square feet each unless you raise them in close confinement which leads to problems. In the same 6 square feet, you can raise 6 quail comfortably.

Since 3 quail eggs are equivalent to 1 chicken egg, you will double your egg output in the same space.

#3 Food Production

Quail eggs taste great and are ⅔ yolk. Each quail will lay about 250 eggs per year. They are full-grown at 4 weeks and begin laying at 6 weeks. That means you could start today and have eggs before summer ends. This is much better for food production than most poultry which takes a minimum of 20 weeks, usually more.

Quail are also very efficient for meat production. They are ready to process in just 4 weeks and dress
out at ⅔ LB each.

#4 Money and Self Sufficiency

Start-up costs for raising quail are lower than most other poultry. One 50Lb bag of feed will feed 9 quail for one year. That means your per-cost egg can be as low as 2 cents. With such small feed requirements, you can easily grow your own feed even on a small homestead. Since quail are easy to breed and raise, you can breed, raise, and easily butcher your own quail at home, making it a very self-sufficient project.

Quail coops are very easy to build because they do not need much space and can be raised in tractors to increase efficiency.

All of these reasons also make quail a great option for a first homestead business. I hope you get quail, I promise you won't be disappointed.

Francis Roland is a young teen who raises quail and other animals on his family homestead in rural South Dakota. He is on a mission to have everyone raise quail for sustainability and fun.

Homeschooling 101

By Dawnita Fogleman

Can I truly fit the nuts and bolts of home education in one article? Well, considering there is a library's worth of books available on the subject, probably not. Seriously though, it shouldn't be that hard. In considering homeschool, there are a few things you need to readjust in your thinking first.

- Homeschool is not school at home. If you try to make it look the same, you will burn out very quickly.
- Education is not K-12+. The goal should be to teach children how to learn and to love doing it.
- Habits and good character are most important. Academics mean nothing without good, solid habits to reinforce everything learned.

How do I know this? Experience. The hard way!

Education is actually pretty simple.

Children need to be able to:
- Read the Bible and the newspaper
- Figure math well enough to make change
- Research
- Ask for help
- Listen to instruction
- Discern truth and fact from deception
- Respectfully and logically debate

While each state has different laws and some parents have to jump some hoops, many everyday activities can be documented as school time.

Preschool should be spent in play. Lots of dominoes, colorful playdough, bean bags and blocks. Simple books and phonics cards. And learning basic obedience and serving others. This is a time when children need to explore. They should begin helping with housekeeping and gardening.

Grammar school is time for soaking up lots of information. Children are sponges during those elementary years. Information should be full of morals, values, facts and character training. As they take in this information, they should be regurgitating it in letters to grandparents, 4-H talks and other opportunities to share. Nature study should be a big focus.

After the age of twelve, I consider children to be youth. (Never "teenagers"! That word tends toward excusing bad behavior and attitudes.) They begin to reason at this age and need to know why and all those facts and all that information they took in earlier works. Therefore, middle school is simply a time of going back over all the previously learned facts and reasoning who, what, when, where, why and how it all fits together.

High school is preparation for the child to enter the adult world of voting, paying bills and earning their way in the world. This is when they need to read the Constitution, classic literature, financial management, and interest-led, life skills. They should be cementing those basic Biblical principles for life. These young adults should begin to develop their own thoughts, ideas, and opinions.

This may seem oversimplified. The idea is, many different books and resources can be used in the process. Keeping perspective is key. It's not the books and curriculum that make home education. Consistency, flexibility, and adventure make true education.

The Importance of Mentorship

By Wyatt Tate, Youth Contributor

Mentors can teach you a lot of things and help you make good choices.

Do you have people you can look up to in your life? You can have mentors for many different things in life.

You need people in your life that have more wisdom than you. They might be not older than you but have life experience. And sometimes they are older. For me, most of the people in my life that teach me things are older and that's okay.

You want people in your life that you can talk to about things like God-things as well as practical things like advice on certain skills or hobbies. Sometimes you're blessed enough to have people that can do both.

I have a good friend at youth group, Caisey, whom I can talk to about making decisions and who helps me through hard times and grow as a godly person. She encouraged me to get rebaptized last month after I had considered how I wanted to rededicate my life to Jesus.

Many people inspire me to be a better person and you can look for people in your life that can inspire you too.

The Bible tells us, "As iron sharpens iron, so one man sharpens another."

I have two really good godly men that help me with hard decisions. My friend Josh, who just joined the military, was homeschooled like me so he knows what it is like and can help me with that. Then, my friend Mr. Greg gives me good advice to help me listen to my parents and respect them. We often go to lunch together at our favorite restaurant in town. It is important to make time to be with your mentors on a regular basis.

I also have special people that I know are better at hunting, trapping, and fishing and I can ask them questions and learn from them.

People like our friends John and Doug who have taught us about trapping have been very important to my Dad and me because we wouldn't know how to trap without them. Without them, we would not have had a successful trapping season last year.

When you want to learn a special skill you should seek out people who know more about that skill than you do and can teach you what they know.

In 2022 we rode with them and learned by watching and helping. Then this year Dad and I set our own traps and Doug or John would check them with us and give us advice. It is important when you get advice from someone with more experience you listen to it.

That's another good point, even when you're older like my Dad, you should still seek out people that can give you wisdom on different things.

And don't forget those closest to you. My Mom and Dad are great mentors.

Sometimes it is easy to forget the people close to you but ask God to remind you how special the people around you are....and never stop learning.

What I Wish I Would Have Done Differently as a Homeschooled Teen

By Christayla Vassar

I wish I did my teen years differently. My mom homeschooled me throughout my entire schooling. God blessed me by being able to go over-the-road with my family in my childhood, as my dad drove a semi-truck for a living. I visited 47 states (only missing Maine stateside!). I also grew up on a small hobby farm as a teen. We raised chickens and Muscovy ducks. I even owned a horse for a few years! Though so blessed, I do wish I did a few things differently to better set myself up for success as an adult.

Learn to drive

I was not someone who was excited about driving. You want me to look at how many places all at the same time and look forward? You know I only have two eyes, right? Plus, other people driving terrified me. I mean, what if someone hit me? What if I hit someone? What if a pedestrian jumped in front of me?! I overthought it a little bit.

Regardless, I wish I would have studied to get my permit and then my driver's license. It would have made it much easier to see my friends who lived across the state, visit family, and get a job.

Go to the vo-tech or take concurrent classes

Concurrent classes terrified me as a teen. Me, in college classes?! How would I keep up?! Plus, I wasn't even sure I wanted to attend college once I graduated. Would it just waste my time and my parent's money to attend? So I did not attend.

I do wish I attended the local vocational school (vo-tech). I didn't know it was even an option for me! That I would have loved, to get hands-on training in graphic design, IT, welding, or even small engine repair.

The moral of this is, check into ALL your options. Even the ones you're not sure if homeschoolers are "allowed" into. You really might be surprised!

Start my business or get a job

I knew I wanted to be an entrepreneur as a teen. I read about business, virtual assisting, and freelance writing. But I simply didn't feel like someone would hire me if they knew I was a teen.

Living so far from town, getting a job wasn't easy either, especially as I was still "in school." I do wish I at the very least found a job, though, even if not starting out as an entrepreneur.

Travel internationally

I traveled all over the USA as a child, but my heart longed to travel across the world. I had (and still have) a serious case of wanderlust. I had the chance to go to Europe with a group called People to People. It didn't happen, but I wish in the deepest part of my heart that it did. I also had the chance to go to Honduras on a church mission trip in my teens. I wasn't able to go to that one either. It's my biggest regret that I didn't travel internationally as a teen.

Grown out of my shyness

I suffered from painful shyness as a child and teen (stereotypical homeschooler, I know). If I had gotten out and done the preceding, but also had support to fall back on, I may have grown out of it sooner.

Take extra electives

I wish I would have asked my mom to buy me that travel writing course as an elective for my junior or senior year. I wish I would have found online classes to support my interests. It's so important to chase those passions you have and to take the classes. Read those books. Talk to the people who are doing what you want to. Find those who can help you make your dreams a reality. When you take those electives, you never know what doors will open up.

Involved myself more in 4-H

I didn't become involved in 4-H until I was a teen. But did I love it! As someone who loved animals and wanted to learn more about animal care, it was great for me. I wish I would have become more involved and actually taken on more of the challenges. I would have learned so much that would have helped me in adulthood. Don't let your teen years (or your teen's teen years!) pass by. Find those things you want to do and that will prepare you for adulthood. Delve into your passions.

If I Were A Pioneer Mother, Here's How I'd Guide You

By Julie Voth

I was reading 'Pioneer Women' by Joanna L. Stratton again earlier this year. In her book, Joanna shares first-account tales of Kansas pioneer women from the early 1900s.

Throughout the book, I was enamored with the way knowledge was passed down through the generations. Essentially, women taught one another everything about womanhood. Their original understanding came from the women before them, who ultimately learned from the Bible.

This is a stark contrast to our modern social media-driven world. These days, many flock to social media for tips on how to handle different parenting situations. In reality, this can cause more harm than anything.

Social media can deliver confusing, overly simplified, and sometimes downright terrible guidance on how to handle parenting situations. Further, the sheer volume of parenting information can be overwhelming. Women might feel stuck wondering what the right steps are. Worse, they can let situations escalate and break down the delicate mother-child dynamic that serves as the very foundation upon which our children are raised.

I am one of the women who listened to social media far too much and let it guide my motherhood journey in unhealthy ways. It lead to confusion and a constant change in how I handled situations to make sure I was 'doing it right.'

Today, I want to help other mothers like me break free from that cycle and plant their feet on a firmer foundation. To do this, we must go back to the women who came before us and learn how to be a mother without any reliance on or influence from the internet.
I pray this helps you the way it helped me.

Get parenting advice from people who know you

The fastest way to break through the noise of social media is to decide that you will not take guidance from social media. Or you can decide you will not let it set the tone for your motherhood. Before social media, women got their advice from each other. The advice was from women who knew their situation and had been there before. It was also easier to discern good advice from bad advice because you knew the person delivering it.

On social media, it's hard to tell as so much is hidden from us. Instead of relying on social media, turn to someone you personally know and trust to give you sound advice. If you don't know who to turn to, turn to the Bible.

Be slow to try new things – unless radical change is necessary

Something I recognized when I was following too many parenting accounts on social media was my tendency to change my approach to motherhood all too often.

I based my approach to handling situations like teaching lessons or delivering consequences on what I'd watched earlier that day. It was driven by the desire to handle it right, and the lack of confidence in what I was doing. This is something I think many women face.

Children need consistency, though. Constantly trying new things to 'fix' them doesn't work. Our kids don't need to be fixed, they need us to patiently guide them to success. Pick one parenting strategy – ideally one you gain from someone you know – and stick with it. Unless radical change is truly necessary, be patient and trust that your methods are working. Kids will be kids, and it will take time, but things will always get better.

Do not compare yourself to other mothers

Social media is a grand tool for comparison. And comparison is a grand way to make you feel ashamed of yourself. Stop.

Instead of comparing yourself to other mothers, focus on being the best mother you can be. Recognize the resources you have. Let God shape your character into one of a great mother and trust in the process. None of us are on the same journey, and women on social media do not set the standard for what "good mothers" are. Social media should never be used to validate that you are a good or bad mother. It can be used for entertainment and inspiration, certainly, but be careful not to cross the line into using it to form your identity.

Your identity is God-given. No one on social media determines it for you. Period.

Love the children God gave you

Ever since Adam and Eve, parents have had expectations for their children. We want them to carry on our legacies and preserve them for future generations. To this day, we still dream of what – or who – our children will be.

While it's wonderful to have dreams for your children, it's unhealthy if those dreams encourage you to push your child to be anyone other than who they are.

You need to learn to love the children God gave you, help them reach their fullest potential in who they are, and let the journey humble you.

You'll be a better mother and woman because of it.

Avoid all-or-nothing parenting extremes

Social media thrives on polarities. Strong viewpoints equal more engagement. This does not, however, equal greater truth. Sadly, too many women do not recognize the hidden agenda of many creators online.

Don't get me wrong: there are incredible, ethical creators, too. As a sort of 'Instagram influencer' myself (if I can call myself that?) I know many of us are striving to put out sound, helpful guidance for other women.

This not the case for all creators. The result is that a lot of what you see is shared for likes and engagement. And, this content can be compelling too because it dances on our emotions of wanting to be the best mothers for our children.

Keep this in mind when following anyone online. You will find some good creators who speak to your unique situations and are ethical in their sharing. But you need to be prudent in weeding them out and avoid following too many so you aren't overwhelmed by their different opinions. This will help prevent the information overload social media can create.

Motherhood is a challenging journey. It is overwhelming and all-consuming, no matter how 'good' we are at it. When we add social media, it is easy to get thrown off track and feel like the worst mom in the world – and like we need to 'fix' that.

The truth is, we all make mistakes. When we build a solid foundation on God's creation for mothers and the sound advice of other women who have gone before us, we can find peace and confidence on the journey.

Current mothers are being led astray in many confusing and contradictory directions because of the sheer volume of overwhelming information available on social media. And while it might seem helpful, in many cases, it's not a truly valuable or necessary resource when parenting.

In many cases, just doing it the old-fashioned way is best. In my case, throwing my motherhood in reverse and running back to the 'old ways' was the best thing I've ever done for myself. I pray my testimony helps you find relief, peace, and confidence, too!

Julie Voth of Mrs. Prairie Wife is a Christian homemaker blogging about faith, femininity, and family. In 2017, Julie and her husband relocated to the Canadian prairies and have been passionately laying roots for their family ever since. You can find Julie at mrsprairiewife.substack.com or @mrsprairiewife on Instagram, Threads, Twitter, and Facebook.

Lone Wolf Prepper

By Steven White

"I am SPEED!" Everyone remembers Lightning McQueen from the Pixar Movie, CARS, and his self-talk. Now substitute "I am the lone wolf!"

A lot of preppers' self-talk is similar:
- "I'm 6 foot, 203 pounds of twisted steel, martial arts master, with a shotgun, rifle, and a four-wheel drive."
- Head nod to Charlie Daniels. "I have more guns than I could ever carry, 10,000 rounds for each."
- "I have 1000 gallons of stored fuel, enough propane to survive and heat my home for over a year..."
- "I can throw knives, hunt and kill anything in North America..."
- "I started my own garden in the past, and know how to grow my own food..."
- "I've got 100,000 calories of canned food stored up in my cellar... and no can opener!"

Yes, Lone Wolf, you have a major flaw in your preps. Everyone has them, and no one is going to be as prepared as they think they are. By the way, Lone Wolf, you're closer to 230 pounds now, 10 years older, your 1000 gallons of fuel, is more like 100 of usable gallons, and your martial arts mastery was some self-help videos on YouTube. Hunting is going to be non-existent in very short order. Yes, there was a flock of turkeys - about 20-25 deep - west of town. But do you think 24 other people aren't going to remember them too as soon as the grocery store lights go out and the freezers fail? So what do you do?

The realization about the "Lone Wolf" mentality, failure often causes stress and depression.

"But I don't trust other people," you cry out. Guess what, they don't trust you either.

"They're not squared away," you cry out. You're not either, remember you forgot the can opener, or whatever structural failure your own preps have.

You know there is something, and you've found and rectified others. You need a "community"!

Now, acknowledging the root word of community leads to commune, or communism, is usually how the lone wolf mentality foments.

No, we're not talking about Soviet-style communism leading to death at the barrel of a gun. Communes have been around since the dawn of recorded time. People getting together to fight a common enemy, starvation, attackers, or weather events.

No one accuses the wagon train drivers who circled the wagons during an attack of being a hippie and living as a communist. They had a common goal, IE survival, and a common plan.

You WILL need other people in times of Grid Down, or services like electricity and government are down. You will need a can opener, more fuel, and someone to watch your child while you procure resources.

The "Stone Soup" story comes to mind.

> The story goes, during a famine, the father had no food for his children, so he was desperate for food for his children. He put some stones deep into a cauldron, covered with water, and set the cauldron to a slow simmer. The children were excited to relieve their pinched bellies, and the smoke from the chimney caused a stir from the rest of the impoverished community. The first neighbor stopped by, inquired about the stew, and the father, indicated surreptitiously, there was no real food there and requested the neighbor play it up for the kids. The neighbor dipped the ladle, sipped the "soup", smacked his lips, and replied, "needs some carrots". Another neighbor came and similarly stated the "soup" needed potatoes. Multiple neighbors became involved, and before you know it, a hearty potato soup with beef, carrots, and all the fixings, became the centerpiece for a neighborhood-wide feast where everyone was satiated. The last person to dip out the actual good soup found the stones, and wondered why they were there, as he was so far removed from the origins of the "Stone Soup" he didn't know the initiation of this blessing, he had only brought a little bit of pepper, as that was all he had. The moral of the story, you may not have all the ingredients, but you may know someone, or several someone's, who might.

The community, YOU need to develop, will have all kinds. Lone Wolves will never think about the 65-year-old grandmother for their prepper community. But, who is going to keep your clothes hemmed up as you lose weight during extended grid-down scenarios? Who is going to bandage up scraped knuckles, or have enough thread to stitch up your fresh gear tactical BDUs? Do you have an extra button? I don't, but I remember my grandmother having an entire gallon jar of buttons that could be used in a time of need. Do you have a gallon jar of buttons? She may just bring the pepper, but you might help her survive the grid down too.

Being perfectly prepped is never going to happen. Not for one year, two, or 10 years. A community will need to eventually come back.

Lone wolf might be good for a week or three days, but what about those around you. Are they liable to not be as squared away as you are? Possibly, yes. What if they're not, what if you are the weakest link, and they were all formerly in emergency management and have been prepping for decades. Do you want them to have the Lone Wolf mindset, and not include you in their reindeer games? No you don't.

Get to know your neighbors. Find out where they are in their preps. Get them up to speed, care about them before SHTF (Stuff Hits the Fan, I don't know what you were thinking), and keep everyone in your group trained and prepared for all potential hazards. Remember if you can only bring the pepper, be glad for it, and thank God someone can bring the potatoes and carrots.

God wants us to love our neighbors as ourselves. That doesn't let us have them freeze in the winter, bake in the summer, or starve while the horn of plenty is abundant at our house. They may only bring some thread to the equation, or it might be the old pumper that knows where to get high octane jet fuel drip you can run your vehicle on with one pass through a coffee filter.

Knowledge and not necessarily stuff is what makes you prepared. Love your neighbors and make some "stone soup" this weekend.

Steven White currently works for the Oklahoma Department of Transportation after spending almost 20 years in Kansas and Oklahoma Law Enforcement and Emergency Management. He has been adhering to the prepping lifestyle and enjoying the prepping tools for approximately 5 years now. Steven enjoys driving to beautiful locations with his wonderful wife, and spending quality time with his children. Join him in his adventures on YouTube and Instagram @OkieDokieExplorers and Twitter @OkieExplorers.

You Can Grow Food… ANYWHERE!

By Christayla Vassar

No matter your living situation, you can grow food. It doesn't matter if you live in a tiny apartment or a sprawling acreage. You can grow food ANYWHERE.

Maybe you think "I can't grow food. I have a black thumb." I firmly believe that even black thumbs can grow food. It's a matter of prioritizing, scheduling, and learning. God's given you the gift of stewardship, and that means even growing plants.

How can you grow food anywhere?

There are several different methods.

Say you live in a small space such as an apartment, townhouse, or tiny home without a yard. You can choose the option of growing using grow lights or a sunny windowsill. No, growing corn or watermelons in your apartment might not work. But you can grow herbs, tomatoes, microgreens, mushrooms, and even crookneck squash!

You can also add other organic matter to your garden. This can work with pots, an in-ground garden plot, and raised beds. These include:

- Wood chips or shavings
- Shredded newspaper
- Leaves, hay, or straw
- Banana peels or eggshells
- Rabbit droppings
- Brown paper bags
- Corn husks
- Hot composted manure (chicken, rabbit, or horse)

Growing in Containers

To grow in a pot, you'll need:

- A pot or planter and saucer
- Sticks
- Organic matter
- Organic potting soil
- Organic compost

First, you'll need a pot. Many things can be used as pots. Here's a list to get you started:

- Mineral feed tubs
- Plastic containers like sour cream and ice cream tubs
- Woven Baskets
- Old stock tanks
- Plastic swimming pools
- Half of a barrel (plastic or wooden)
- Burlap bag
- Laundry baskets
- Grow bags (you can find economical ones!)
- Milk jugs
- Old cake pans
- Trash bags
- Food grade buckets
- And anything else you can find!

1. When you have your pot, drill any holes in it that you'll need for drainage.
2. Fill the pot with a layer of sticks. This will not only take up room in your pot but allow room for the plant's roots to breathe. They also release nutrients into the soil as they break down.
3. Next, add some organic matter. Then add your organic potting soil and compost. You can plant your seeds directly into this.

Shady Garden

Maybe you have a shady part of your yard you want to plant in. But you don't think you can because it's so shady. There are several varieties of plants you can plant in the shade.

Depending on the amount of sunlight your space gets, you can plant the following:

Tolerate 3-4 hours of sunlight:
- Salad greens such as lettuce, spinach, arugula, sorrel, and endive
- Leafy greens such as Swiss chard, mustard greens, kale, and collard greens

Tolerate 4-6 hours of sunlight:
- Onions, shallots, and leeks
- Potatoes
- Beets
- Carrots
- Radishes
- Rutabaga
- Turnips
- Cabbage
- Brussel sprouts
- Broccoli
- Cauliflower
- Peas
- Beans
- Parsley
- Cilantro/coriander
- Oregano
- Chives
- Mint

These plants may grow slower or smaller than their counterparts in full sun.

Raised Beds in the Cheap

It's cheaper than you may think to make a raised bed. You can make them out of pallet wood or used lumber. You can also buy kits to make corrugated metal raised beds. You can even use landscape poles, large logs, or wattle with stakes to keep them together.

After you've made your raised bed, it's time to fill it. You'll need:

- Hardware cloth (optional)
- Cardboard or newspapers
- Logs and sticks
- Organic matter

- Unsprayed Topsoil
- Organic manure
- Coconut coir or vermiculite (optional)
- Organic compost

1. Optional step: Put down a layer of hardware cloth if you live in an area with moles and gophers. This will keep them from disrupting your garden and stealing your root veggies.
2. Start with cardboard or newspapers for your first layer. This will kill any weeds underneath. Water this layer.
3. Next, add a layer of logs and sticks. Depending on the depth of your raised bed, you can fill the raised bed halfway to three-fourths of the way to the top.
4. Add organic matter, like leaves, straw, and shredded newspaper. You can add veggie scraps such as banana peels and egg shells here too.
5. Next, mix topsoil and manure at a ratio of 4:1. Too much manure can burn your plants. Add this mixture to the pile. You can also add coconut coir or vermiculite for airier soil.
6. Finish the raised bed with a layer of compost. Place straw on top to keep the compost from blowing away.

There you have it! A new raised bed.

In-Ground Garden

A no-till lasagna garden doesn't require a raised bed. You can find different methods to make an in-ground garden, but I'm going to show you an easy one.

You'll need:

- Cardboard or newspapers
- Sticks
- Organic matter
- Unsprayed Topsoil
- Organic Manure
- Organic compost
- Unsprayed straw
- Coconut coir or vermiculite (optional)

1. Place the cardboard or newspaper down first on your site. Add sticks on top to keep it from blowing off. Water this layer.
2. Next add organic matter such as straw, coconut coir, wood chips, and leaves. Water again.
3. Mix topsoil and organic manure 4:1 and add to this. You can also add coconut coir or vermiculite to this mixture. Water again.
4. On top add about an inch of organic compost.
5. Keep the soil covered with straw until you're ready to plant.

Hugelkultur

Do you have a bunch of logs or fallen branches lying around? Then you might want to make a hugelkultur bed.

You'll need:

- Cardboard or newspapers
- Logs and sticks
- Organic matter
- Unsprayed topsoil
- Organic compost

1. Dig down a few inches and remove the sod.
2. Place cardboard or newspapers down to kill off any grass or weeds seeds.
3. Stack your logs and sticks until they are 4-8 feet high. Remember, a hugel mound WILL shrink over time.
4. Add your organic matter on top. You can add topsoil too.
5. Add organic compost on top.
6. Optional: On top of this you can sow clover seed if you aren't going to use it right away.

You've completed your hugelkultur mound!

Just the Facts - Politics from the horses' mouth…

By Dawnita Fogleman

Agenda 2030 is officially in full swing worldwide this summer. NASA has announced an intention to launch two missions to Venus by 2030, "to study the greenhouse effect and learn how to manage it on Earth," according to The Evergreen Aviation and Space Museum.
https://www.evergreenmuseum.org/2021/06/06/fa-18-super-hornet-sky-jump-launch-3/

In June, King Charles III attended the Climate Innovation Forum to launch a Climate Clock, warning there are only six years and 24 days left to limit global warming to 1.5 degrees, according to Climate Action.
https://www.climateaction.org/news/king-attends-national-climate-clock-switch-on-at-climate-innovation-forum

In July, Biden met with the new King announcing key players and investors for "Climate Finance."
https://www.whitehouse.gov/briefing-room/statements-releases/2023/07/10/joint-fact-sheet-president-biden-and-his-majesty-king-charles-iii-meet-with-leading-philanthropists-and-financiers-to-catalyze-climate-finance/

The office of the Director of National Intelligence launched a program to develop new innovations for tackling threats and advances inherent within the rapidly changing biointelligence and biosecurity landscapes.
https://www.dni.gov/index.php/newsroom/press-releases/press-releases-2023/item/2380-iarpa-pursuing-breakthrough-biointelligence-and-biosecurity-innovations

The White House announced more than $8 billion in new commitments as part of call to action for ending hunger and reducing diet-related disease by 2030.
https://www.whitehouse.gov/briefing-room/statements-releases/2022/09/28/fact-sheet-the-biden-harris-administration-announces-more-than-8-billion-in-new-commitments-as-part-of-call-to-action-for-white-house-conference-on-hunger-nutrition-and-health/

WHO staff convened in Geneva about the implementation of Pandemic Influenza Preparedness (PIP) Framework activities under the newly published High-Level Implementation Plan III (HLIP III) for 2024-2030.
https://www.who.int/news/item/14-07-2023-realising-the-next-pip-partnership-contribution-high-level-implementation-plan-(hlip)-for-2024-2030

U.S. Congressmen introduced The Social Media Child Protection Act and Kids Online Safety Act making it the social media platform's responsibility to verify age using methods like ID verification.
H.R.821 https://www.congress.gov/bill/118th-congress/house-bill/821/text
S.1409 https://www.congress.gov/bill/118th-congress/senate-bill/1409/text

Passing on Traditions

Fall 2023 · PrairiDustTrail.com · Volume 01 Issue 04

PrairieDustTrail.com
Connect with the past ~ Prepare for the future ~ Live more sustainably

Passing on Traditions

Traditional
- Skills
- Gifts
- Mindset

Editor's Epistle

Daniel answered and said, Blessed be the name of God forever and ever: for wisdom and might are his: And he changes times and the seasons: he removes kings, and sets up kings: he gives wisdom to the wise, and knowledge to them that know understanding: He reveals the deep and secret things: he knows what is in the darkness, and the light dwells with him. Daniel 2:20-22

Fall is my favorite time of year, mostly because Oklahoma is brutal in summer. As autumn moves in, there will still be a few 90+ days, but they'll be further and further apart as the weather is drug, kicking and screaming into winter. Through the winter, things slow down with the worst of the chores being breaking the ice in the stock tanks. Yes, we've invested in heaters before, but even they will last longer if the ice stays broken up as much as possible. Then we can look forward to spring and everything hopefully coming to life again with rain, the stirring of the winds, and lightening ionizing, freshening the air. Three whole seasons without 100+ degree dry wind is such a refreshing blessing!

Isn't this the way things are in life as well? In marriage, we may be looking forward to the spicy hot passion of a summer season, while winter's cold shoulder is least enjoyable. With children, the calm of fall gives us a reprieve from the spring storms. Our personal and spiritual growth is even a tumultuous bout of seasonal tossing around.

As we move into the season of Thanksgiving and gift-giving, I pray we are all mindful and intentional about each day, every activity, and all of our relationships.

Faith Has Made.com

· Handmade turquoise jewelry
· Product photography
· Learn to make jewelry

by Abagail Gray
@faithhasmade on FB & IG

Breaking the Curse of Generational Trauma: A Path to Healing

Kathryn White

Exodus 20:5 - ...for I the LORD thy God am a jealous God, visiting the iniquity of the fathers upon the children unto the third and fourth generation of them that hate me...

Generational trauma, also known as ancestral or intergenerational trauma, refers to the emotional, psychological, and even physical scars that can be passed down from one generation to the next. These wounds are often the result of traumatic experiences, such as war, abuse, addiction, discrimination, or even silent sin, and can cast a long shadow over a family's future. However, the cycle can be broken.

Understanding Generational Trauma

Generational trauma is a complex phenomenon, often rooted in a family's history. These traumas can manifest as patterns of behavior, beliefs, and emotions that persist across generations. For instance, a family with a history of infidelity may find that self-sabotaging behaviors are passed down through generations. Or, a family with a history of passive aggressive behavior may havBreaking the Cycle.

Breaking the cycle of generational trauma is challenging, but essential, especially if we want our family to be salt and light. Here are some key steps to help individuals and families heal:e a harder time setting boundaries and being genuine with others. This cycle can perpetuate suffering, making it essential to recognize and address the issue.

The Impact of Generational Trauma

Generational trauma can have a deep impact on individuals and families. It can lead to various mental health issues, including anxiety, depression, and post-traumatic stress disorder (PTSD). It can also affect relationships, leading to dysfunctional family dynamics, strained relationships, emotional distance, and divorce. All of those issues then go on to create other damaging cycles.

Acknowledge the Trauma:

James 5:16 - Confess your faults one to another, and pray one for another, that ye may be healed. The effectual fervent prayer of a righteous man availeth much.

The first step in breaking the cycle is to acknowledge the presence of generational trauma within your family. This requires an honest examination of you and your family history, recognizing the patterns and behaviors that have been passed down, and the patterns and behaviors that YOU have adopted and employed.

Seek Professional Help:

Proverbs 11:14 - Where there is no guidance, a people falls, but in an abundance of counselors there is safety.

Therapy or counseling can be a crucial component of healing from generational trauma. A mental health professional can provide guidance, support, and tools to help you and your family address the trauma and develop healthier coping mechanisms.

Build Awareness:

Deuteronomy 28:1-68 - And it shall come to pass, if thou shalt hearken diligently unto the voice of the LORD thy God, to observe and to do all his commandments which I command thee this day, that the LORD thy God will set thee on high above all nations of the earth...

Understanding the root causes of generational trauma can help you become more aware of how these patterns manifest in your life. Awareness is the first step toward change. A therapist or good counselor should help you start growing that awareness, but it will take work from you more than anything; being curious about why you and your family act, react, and behave in certain ways, and you gently digging to find the root cause.

Break the Silence:

John 8:32 - And ye shall know the truth, and the truth shall make you free.

Generational trauma is often accompanied by secrecy and shame. Breaking the silence within your family and openly discussing the trauma can be empowering. It allows family members to connect, share experiences, and work together to heal.

Practice Self-Care:

Galatians 6:1 - Brothers, if anyone is caught in any transgression, you who are spiritual should restore him in a spirit of gentleness. Keep watch on yourself, lest you too be tempted.

Taking care of your own well-being is essential. Self-care can involve mindfulness practices, physical activity, healthy eating, and setting boundaries to protect your mental health. This is crucial for you to become more aware and emotionally healthy enough to continue healing.

Educate Yourself:

Proverbs 16:16 - How much better to get wisdom than gold! To get understanding is to be chosen rather than silver.

Learning about generational trauma and its effects can be empowering. There are numerous resources available that can provide valuable insights and strategies for healing. Start by reading books (I suggest anything by Dr Caroline Leaf and Dr Henry Cloud), following accounts on social media (I recommend the Holistic Psychologist and Attachment Nerd), and of course joining groups and talking to other people on their healing journey who have experience.

Establish Healthy Relationships:

1 Thessalonians 5:11 - Therefore encourage one another and build each other up....

Building healthy relationships with friends and loved ones can provide a strong support network. Surrounding yourself with positive influences and healthy boundaries can counterbalance the negative impact of generational trauma.

Break the Generational Patterns:

Romans 12:2 - Do not be conformed to this world, but be transformed by the renewal of your mind, that by testing you may discern what is the will of God, what is good and acceptable and perfect.

Consciously choose to break the generational patterns that have been passed down. This may involve reevaluating your own behaviors and beliefs and making a conscious effort to change.

Encourage Family Healing:

Hebrews 10:24-25 - And let us consider how we may spur one another on toward love and good deeds, not giving up meeting together, as some are in the habit of doing, but encouraging one another—and all the more as you see the Day approaching.

If possible, involve other family members in the healing process. By working together, you can create a stronger support system for everyone.

Practice Forgiveness:

Colossians 3:13 - Bear with each other and forgive one another if any of you has a grievance against someone. Forgive as the Lord forgave you.

Forgiveness does not mean condoning or forgetting the trauma, but it can help release the emotional burden of generational trauma. Forgiving those who caused the trauma, including oneself, can be a powerful step towards healing.

Healing Is Rewarding

Generational trauma - curses - is a heavy burden to bear, but it is not an inescapable fate. By acknowledging the presence of trauma, seeking help, and taking proactive steps to heal, you and your family can break the cycle and move towards a healthier, happier future. While the journey may be very challenging, the rewards of breaking free from generational trauma are immeasurable – a legacy of resilience, strength, and healing for generations to come.

Daniel 4:34 ...and I blessed the most High, and I praised and honoured him that liveth for ever, whose dominion is an everlasting dominion, and his kingdom is from generation to generation...

Christian Herbalism

By Kristen Smith

Herbal medicine is one of the oldest forms of healing. People have been using plants to stay healthy, treat sicknesses, and enjoy life for thousands of years. But about a century ago, for a number of reasons, the American herbal tradition died. The culture shifted to what we now call modern medicine and turned its back on traditional remedies. This wasn't all bad. After all, antibiotics and surgeries have proved life-saving in many situations. Yet we're now left with a culture that requires pills for every minor concern, can't identify any medicinal plants, and sprays healing weeds with glyphosate. It's not healthier, either.

Thankfully, herbal medicine is making a come back. More and more people are seeing the potential harms in popping too many pills and want to work with nature to feel their best.

What is Christian Herbalism?

Most herbal practitioners and teachers aren't Christians, though. You'll find everything from shamanism, New Age, agnosticism, Eastern religions and philosophies, atheism, Wicca, Druidism, pantheism, and nothing or everything in between. These people are often highly skilled at using herbal remedies. However, they may not be friendly towards the Christian faith.

There's a small but growing group of Christian herbalists, though. I'm one of them.
We strive to know and use our herbs well. We desire to help our students, readers, and clients experience better health through plant medicine. And we want the true God of the Bible to get the glory for it all.

Why the Faith Element Matters

I recently wrote to my email community about a popular and engaging herbal teacher making the rounds in the mainstream herbal community. He was marketing his herbal program, which no doubt offers a lot of insightful, well-rounded instruction. Yet he weaves alchemy, astrology, and achieving a higher spiritual consciousness all through
his teaching. It's sadly reminiscent of "...ye shall be as gods" from Genesis 3:5, seeking wisdom and divinity without the true source of such things.

And so I felt compelled to write to them with a gentle warning about why I could never encourage an herbal program like that.

Satan is wise and cunning, "...seeking whom he may devour..." as 1 Peter 5:8 says. He'll use anything, even lessons on herbal remedies, to get us off the straight and narrow way that leads to eternal life with Christ. Because this kind of herbal teaching is the norm in mainstream herbalism, carrying on a Christian herbal tradition has become one of my purposes in this life.

Passing on the Healing Tradition

I love helping people see how herbal remedies and faith in Jesus Christ go hand in hand. After all, Christ is "…before all things, and by him all things consist," according to Colossians 1:17. This world and all that's in it was created by Him, is held in order by Him, and is here to bring Him glory. Even medicinal plants!

Scripture also teaches us that "He causes the grass to grow for the cattle, and herb for the service of man…" in Psalm 104:14. Life-giving plants, whether they be broccoli, apples, or echinacea, were created by the Lord for our health and benefit. When I harvest medicinal herbs at my home, host Seeking Whole Health, my Christian natural health conference, or write a new article on my website, my goal is to pass along a Christian herbal tradition to those around me.

But what if you're not an herbalist? What if you're a busy homesteader who likes using natural remedies but isn't an expert? There are many ways you can foster and pass along a Godhonoring herbal tradition in your family and community!

- Learn about the wild plants growing on your property. They have names, and often, they have medicinal or food benefits, too!
- Grow a few medicinal herbs, even if you don't get around to harvesting them. Simply having them around will attract pollinators and help you appreciate God's creativity.
- Host an herb-walk at your home or somewhere else. If you don't know enough to lead it, find a local herbalist or forager who can.
- Try making a simple remedy at home. Elderberry syrup is a great place to start, along with plantain salves and echinacea tincture.
- Tell the children in your life about God's provision through healing plants. Help the next generation appreciate strong physical health for God's glory!

The Christian herbal community may be small, but we're growing. And we'd love to have you join us as we help more people embrace a Christian herbal tradition.

Kristen Smith is an herbalist and aromatherapist with multiple natural health certifications. She believes natural health is a blessing from God that you can start enjoying right now, one simple step at a time. You can find her at A Better Way To Thrive (https://abetterwaytothrive.com/) where she shares free articles, publishes helpful resources, and works one-on-one with her Thriving Health clients.

My Window into the Past - Passing on Traditions

By Myrna Buckles

There are many ways we pass traditions from one generation to another.

Traditions are important for a multitude of reasons, some are:
- Experience equals wisdom
- Learning what works well and what doesn't
- Knowing our traditions and history can prevent us from repeating mistakes of the past.
- We want to know the "road out of the wilderness" - remember the Israelites wandering in the desert for all those years?
- We learn skills through traditions.

Historically, passing on traditions has been accomplished throughout the ages via oral storytelling, written history, gift giving, lifestyle, shared knowledge, mentoring, a variety of art forms, needlecrafts, card making, scrapbooking, collections, photography, recipes, ministry, missionary work, trades, etc.

The tradition of handmade gifts was prevalent during the 1920's, 30's & 40's when money was tight. Handmade cards were one of the most frequent handmade gifts. Many adopt the tradition of making gifts as a way of giving a gift with meaning.

When my kids were in grade school, I set aside a weekend for crafting and making gifts for teachers, friends and family. Since my husband typically attended a church men's retreat in the fall, it was the perfect 3-day weekend to make handmade gifts. We made Santa's out of pop cans, paint and felt; snowmen out of socks, buttons and dried beans; and other simple gifts.

Storytelling and Traditions

Storytelling whether oral or written form has been practiced for centuries. The Bible is an amazing example of written stories and traditions and the inspired word of God. There is a reason God made sure it was recorded.

I grew up listening to my dad's stories of his childhood. In later years, we revisited those stories and recorded them digitally so I could preserve them for future generations. Those stories are priceless and give us a view into the past.

In recent years, I have had the desire to learn more about my parents growing-up years in an effort to understand them better. It has been a personal quest as I have learned more and more about how our past impacts our present decisions. The past is a great teacher and we need to delve into it to grow wiser. I'm not saying to camp there, simply visit the past to learn from it.

In an effort to learn how others experienced traditions being passed on, I reached out to friends, family and social media.

Below are some of the responses I received regarding gifting:

- Making special food gifts (homemade vanilla, harvesting honey and gifting it, gift baskets with homemade goods).
- Gather age appropriate books over several months in order to give one per day to each child in the family for the first 25 days of December.
- In place of gifts, donate items to special holiday blessings at church for those in need.
- Give Christmas jammies and a themed book to read together on Christmas Eve.
- Making inexpensive gifts with your children for them to give.
- Making inexpensive and simple table runners. This youtube video describes how to make 10 minute table runners. We all need things that don't take long. I have made this table runner multiple times. It is quick, easy and you alter it to your liking. https://youtu.be/oi4Qgq0KKJQ?si=FuADLRl_WElre_Ra

I also received these responses about traditions that have been passed on by parents/grandparents. These are the responses I received:

- How valuable thoughtfulness & kindness is in life.
- To be kind to others.
- Godly Love.
- To read & pray.
- How to cook chicken & dumplings.
- How to print church bulletins on a big copy machine.
- How to bake, sew, clean, & listen.
- Faith and prayer.
- How to make homemade meatloaf.
- How to ride a horse
- How to sew
- How to make quilts.
- How to garden.
- How to cook, sew, clean and go to church.
- How to be a gracious hostess to whoever entered my home.
- Playing card or board games, collecting, and writing stories.
- How to make potato lefse.
- How to garden, can, cook recipes, baking pies, bread,etc.
- How to make sauerkraut.
- How to homeschool.

My challenge for each of you is to make a list of traditions you have learned from your parents, grandparents and others. Then, make a list of traditions you want to pass on to your kids and grandkids. Maybe it will include starting new traditions and it might include eliminating some traditions.

The Folksy Side of Life

JOURNEYING TO FIND AND SHARE ANCIENT WISDOM IN OUR MODERN WORLD.

www.thefolksysideoflife.com

Sourcing Homestead Items Cheaply

By Christayla Vassar

Homesteading can cost a lot both in time and money. No matter where you homestead, you can source the items for cheap or even for free! You simply need to keep your eyes peeled.

Ask your neighbors

Sometimes your neighbors will start plants, breed chicks, or have a wood shop. They may sell these to you for cheaper than a store-bought plant would be. Or they may know someone who would.

Tap your network

Your network is a great place to find homesteading items. Maybe your father-in-law knows someone in construction who has a dumpster full of wood off-cuts. Or maybe your friend's aunt has rabbit cages she needs out of her garage. You never know what you may find.

Craigslist

Craigslist can be a great place to find free or super cheap homesteading items. They have a free section that can be full of items like stones, fencing, and even pallets. Use caution. Craigslist is also known to have some scams.

Facebook Marketplace or Groups

The Facebook marketplace can be another great place to find homesteading items. It's less scammy than Craigslist but still do your due diligence. You never know what you'll find. You can find partial boxes of linoleum for your chicken coop for a steal! Or free fencing from someone who needs it gone yesterday! You can even find free wood.

Facebook groups, especially local ones, can steer you in the right direction regarding where to source things locally. Who knows... you may also make some friends that way!

Side of the Road

You may find lots of roadside treasures depending on the part of the country you're in. I've seen wood, fencing, tires, pallets, and more left on the side of the road. Find a place to pull over and stop to safely to collect these.

Dump

In some places, it's easy to access the dump. You can find many items here. It is as they say: one man's trash is another man's treasure.

Swap Meets and Flea Markets

Swap meets and flea markets are great places to find homestead items. They can be harder places to find rurally, but it's worth it if you do. You can find fencing, livestock, and animal care items at swap meets in particular.

During Free Junk Pickup

Have you ever seen a bunch of junk piled at the edge of the curb when you go into town? There's a good chance that the city or town is hosting a free junk pick-up event. This is an excellent time and place to find wood, pots, and even garden pavers. My grandma-in-law once found me a folding laundry rack this way!

Garage/Yard/Estate/Moving Sales

When people have too much in their lives, it gets put in a sale. You can find so many things here for great prices. From fencing and posts to flower pots and buckets, a sale is a great place to find what you need on your homestead.

So what are you waiting for? Get your budget together and go out to find your items on the cheap!

Passing on Traditions

By Andrea Hutchison

My great-grandfather Oscar Chain homesteaded the land I call home in 1893. He traded a shotgun and $50.00 for it. He was 17 at the time and too young to file a claim, so made the deal later with a previous homesteader whose wife got homesick. My thought is this poor lady had no idea what she was getting herself into when her husband said, "I have an idea, let's move to a land of blackjack trees, sandburs, coyotes and no neighbors for miles...you'll love it." The poor lady didn't.

Pioneer Oscar Chain (1875-1954) came to Oklahoma with a wagon, team of mules, and $35 to become one of the best known and wealthiest ranchers in Northwest Oklahoma.

I'm glad my grandfather made the trade. Living on this land has created who I am.

A few years after the trade Oscar would marry Laura Hickock (yes Wild Bill's cousin), and to that union came my granddad, Lenard. Laura was big and strong and loved chickens, turkeys and her garden. She would cook for many harvest crews during her lifetime. I share many of those loves. I would be satisfied if I never went to town and could hang with my animals all day. I've cooked for many a hand as well, from cattle gatherings to harvest crews. Her equipment was a cast iron stove and wood. Mine total electric or gas. I don't believe I would trade.

My mom, Darla, was a city girl. She was a great cook, a wonderful artist and loved nice things. My grandmothers, Grace Chain, Lenard's wife, and Hazel England, my mom's mother, were good cooks, devoted to their families and communities.

My grandmother Hazel was a partner in the Stong and England Grocery with her husband Les England and her brother Harold Stong, for five decades. My memory of her is "doing books" in a tiny room, her "office" in the back of the store, or up front checking out customers. Her motto was, "the customer is always right".

My aunt Wymola (Chain) Sander was a powerful influence as well, she kept me plugged into town life making sure I got to the Canton Christian Church VBS and caught the bus to Fairview for swimming lessons, Canton had no swimming pool back then. When I was 8 she was killed in a plane crash along with her husband and another couple. That impacted my life forever.

With the homesteading on every 160 acres, the landscape from Oscar's time to later years changed drastically. Country schools popped up along with women's clubs. Consolidated VI was a large country school located about six miles from my house.

A women's extension club known as Con Six Club emerged in the early 1900's and was the hub of this growing community. The ladies from around the community would gather monthly to discuss the upkeep of the outhouse at the local cemetery or what was needed for upcoming holiday festivities.

Some of the greatest memories I have are attending these meetings and enjoying the local extravaganzas. No jeans or sloppy shirts. The dress code would have been considered business formal I'd say. Sloppy dressing was reserved for chicken killing and laundry day, never for club.

These women were great influencers in my life, but I must be honest most of my mentors were ranch hands. I liked domestic life but loved ranch life more. From fishing, to horseback riding, to showing livestock, I was with one of them or my grandad Lenard and dad, Ralph Chain from dawn until dusk. The women taught me to cook and have a little etiquette but my heart was outdoors.

The most important tradition passed down to me was experiencing the dedicated faithfulness my dad had in making sure we were at church. Sunday morning and evening we were sitting in the pews of the "Y" Church of Christ located three miles from my home. No questions asked.

For as long as I can remember up until my young adult life we rarely missed. I was baptized there and hold many fond memories. From a mouse running back and forth across the glass baptistry to watching a squirrel build its nest in the chimney, I can still see the faces of those who attended and where they sat. Funny what you remember.

So, passing down traditions to me encompasses quite a bit. Community stands out. Faith in God. Structure. Dedication. Faithfulness. Standing by your word. Good Stewardship. My dad always said what we have been given is just on loan from God and we are just here taking care of it. I guess I try to apply that to my family, husband, home kids, health, animals, chores, church work, etc.

I grew up in a very blessed era that included a God-fearing family and community. My goal is to keep that blessing alive. God gives us gifts we can decide to use and share with others, or we can sit and complain. I pray I keep sharing these traditions for as long as I can.

A Republic of Good Behavior vs. A Democracy of Tyrannical Nobles

By Dawnita Fogleman

The United States is a constitutional republic. Democracy leads to tyranny. Why then have even many of our elected officials been throwing around the word "democracy" over the past several decades? Unfortunately, it is not accidental.

Our representative form of government has checks in place to keep the government itself under control. Once a government is under the whim of the populus, an undereducated and uninformed populus at that, it is susceptible to manipulation by the few.

What has gone wrong? Other than the degradation of language as a whole, the decay of education, the warping of our language and vocabulary has been under constant attack.

According to Webster's 1828 Dictionary,

> a republic [Latin respublica; res and publica; public affairs.] is a commonwealth; a state in which the exercise of the sovereign power is lodged in representatives elected by the people. In modern usage, it differs from a democracy or democratic state, in which the people exercise the powers of sovereignty in person.

According to James Madison in Federalist Paper No. 39,

> *"If we resort for a criterion to the different principles on which different forms of government are established, we may define a republic to be, or at least may bestow that name on, a government which derives all its powers directly or indirectly from the great body of the people, and is administered by persons holding their offices during pleasure, for a limited period, or during good behavior. It is ESSENTIAL to such a government that it be derived from the great body of the society, not from an inconsiderable proportion, or a favored class of it; otherwise, a handful of tyrannical nobles, exercising their oppressions by a delegation of their powers, might aspire to the rank of republicans, and claim for their government the honorable title of republic."*

It doesn't take much to see our problem today. Unfortunately, the "people" have continued to elect the same legislators over and over again, regardless that they no longer (and many never did) represent the people they serve, nor do they serve with any limit whatsoever, much less "good behavior" in the slightest. The representatives of the United States of America have become nothing more than tyrannical nobles exercising oppression and blaming one another for the consequences of their irresponsible actions.

Is it any wonder what we're seeing today has been accepted and overlooked for so long? Atrocities against basic, God-given human rights that would have been scorned in so many other countries are now commonplace in what used to be considered the "Land of the Free".

And where is the outrage?

Tucked away behind censorship under the guise of "fact-checking". Today, testimony in Congress is considered false information if it doesn't line up with the narrative fed to an artificial intelligence.

> *"Woe unto them that call evil good, and good evil; that put darkness for light, and light for darkness; that put bitter for sweet, and sweet for bitter! Woe unto them that are wise in their own eyes, and prudent in their own sight! Woe unto them that are mighty to drink wine, and men of strength to mingle strong drink." Isaiah 5:20-22*

So there we are... At this point, I wonder how many people trust anything they see or hear in the news, and I'm not even considering the mainstream news. The rest, I'm afraid, are so desensitized that they don't even seem to care about anything beyond the constant stress of their own little worlds.

At some point, and I'm fairly sure it will be sooner than later, the cold hard truth is going to hit home in a very tangible way. People will (many already are) realize how petty the cares of this world have become.

In the process, for those of us who are awake, or waking, we can stand on the promise of the Church of Philadelphia:

> *"Because thou hast kept the word of my patience, I also will keep thee from the hour of temptation, which shall come upon all the world, to try them that dwell upon the earth. Behold, I come quickly: hold that fast which thou hast, that no man take thy crown. Him that overcometh will I make a pillar in the temple of my God, and he shall go no more out: and I will write upon him the name of my God, and the name of the city of my God, which is new Jerusalem, which cometh down out of heaven from my God: and I will write upon him my new name." Revelation 3:10-12*

3 SUPERFOODS THAT ARE TRULY HEALTHY

by Deborah Hanyon, MPH, RDN, ACE-CHC

There is a lot of hype these days over the idea of superfoods. So, I thought it would be appropriate to discuss some of the foods.

But first a definition. A superfood is defined as "a nutrient-rich food especially beneficial for health and well-being."

Avocados

- Contain two times as much potassium as bananas.
- These delicious fruits are high in monounsaturated fat, which lowers HDL.
- Delicious avocados are also high in folic acid and vitamin B6, two important vitamins that are low in the average person's diet.
- Avocados are also rich in fiber.
- They are high in glutathione, which reduces the risk of some cancers.
- In addition, avocados are high in magnesium which is important for healthy muscles and heart.
- The delicious green gems are also high in Vitamin E, an important fat-soluble vitamin and antioxidant.

Beets

The vibrant color of beets is your first clue that these unique vegetables are highly nutritious. But what is the nutrient behind the color of beets? The nutrients are called "Betalains." Betalains are members of the carotenoid family. Carotenoids are rich in antioxidants. Antioxidants protect against damage from the sun and other environmental toxins. The more vibrant the color, the richer the concentration of phytonutrients in a vegetable or fruit. * And beets are no exception.

Beets, including the greens, are rich sources of about every vitamin you can think of: Vitamin A, C, potassium, folate, B6, iron, manganese, magnesium, thiamin, carotenoids, anthocyanins, to name a few.

> *NOTE: Beets are an excellent source of natural food colorings. Click here for an awesome website that teaches all about natural dyes, including beets.

Apple Cider Vinegar

Apple cider vinegar is the result of the fermentation of apple cider, which is made from fresh, crushed apples, including the cores, peels, and flesh.

- Fermentation occurs when yeast is added to the apple cider mixture. This process produces acetic acid, also known as apple cider vinegar.
- The healthiest form of apple cider vinegar is the unfiltered, unpasteurized form. This is because the enzymes remain active.
- Regardless of whether it is pasteurized or not, however, studies show that the acidity in apple cider vinegar protects the stomach against pathogens.
- The acid stimulates the production of hydrochloric acid in the stomach, which is essential for proper digestion.
- In some studies, apple cider vinegar caused leukemia cell death and inhibited tumor growth.
- Apple Cider vinegar is also a source of polyphenols, protective ingredients found in plants.

Interesting Folklore on Apple Cider Vinegar

1. Hippocrates used vinegar to manage wounds.
2. Hannibal of Carthage used vinegar to dissolve boulders that blocked his army's path.
3. Cleopatra dissolved precious pearls in vinegar and offered her love potion to Anthony.
4. Sung Tse advocated hand washing with sulfur and vinegar to avoid infection during autopsies.

Recipes

by Deborah Hanyon, MPH, RDN, ACE-CHC

Cottage Cheese with Avocado and Salsa

I created this recipe myself years ago when I had cottage cheese in the house and wanted to make it more interesting. This will make a complete meal with a few whole grain crackers or slice of bread on the side.

Prep Time: 10minutes minutes

Servings: 4

Ingredients

- 2 whole Avocados cut in half, pitted and sliced
- 2 cup Low-fat cottage cheese
- 2 tbsp Salsa

Instructions:

- Cut avocados in half; remove pit, peel and slice
- place 1/2 cup cottage cheese on small plate or bowl
- place 1/2 of avocado around the outside rim of cottage cheese
- add salsa to top of cottage cheese

Nutrition

Serving: 0.25Recipe | Calories: 204kcal | Carbohydrates: 10g | Protein: 16g | Fat: 12g | Saturated Fat: 2g | Polyunsaturated Fat: 1g | Monounsaturated Fat: 8g | Trans Fat: 0g | Cholesterol: 20mg | Sodium: 520mg | Potassium: 484mg | Fiber: 5g

Harvard Beets

Fresh or frozen will do. This recipe uses frozen which can be purchased at Trader Joes or other grocery stores.

Servings: 5

Ingredients

- 16 oz Beets frozen
- 1 cup Orange juice

Instructions

- Put frozen beets and 1 cup orange juice into saucepan
- Cook at medium heat until tender, about 10 minutes (add water if needed)

Nutrition

Calories: 61kcal | Carbohydrates: 13g | Protein: 1g | Fat: 0g | Saturated Fat: 0g | Cholesterol: 0mg | Sodium: 71mg | Potassium: 394mg | Fiber: 2g | Sugar: 10g | Vitamin A: 130IU | Vitamin C: 29.2mg | Calcium: 20mg | Iron: 0.8mg

SAUERKRAUT

By Randy Kocourek

Sauerkraut is always better when made in a stone crock. Begin by buying or raising firm solid cabbage heads. Usually all leaves not wrapped tight on the head are removed and discarded. Wash cabbage heads thoroughly with cold water, and then remove 2-3 more leaves from each head and set aside for use later. Depending on size, heads are normally halved or even quartered, and the core removed.

The following ingredients are needed to make the sauerkraut:
- White onion, 3" diameter, one for each 10# of cabbage.
- Caraway seed, one teaspoon for each 10# of cabbage.
- Dill seed, ½ teaspoon for each 10# of cabbage.
- Canning salt, 3½ tablespoons per 5# of cabbage.

It is better to be on the salty side. Too little salt will not allow the fermentation process to work as needed. When canned and ready to eat, the kraut can always be rinsed with clear water to remove some of the salt taste..

Select a crock large enough to handle the cabbage you have. Roughly 25-30# should fit into a 6-8 gallon crock. Shred cabbage on a kraut cutter about 5# at a time. Try to get about 3-4" of depth in the crock. Sprinkle in the caraway, dill seed, and canning salt. Then you must ball up your fists and knead the cabbage, working up and down around the entire crock until water starts to separate from the cabbage. This effort mixes the spices and salt throughout the cabbage to ensure even seasoning. Repeat this process until the crock is filled up to or less than two inches from the top. It is acceptable to skim off some of the juice worked out of the cabbage and discard.

A taste of the juice will tell you about the saltiness of the brine and if you have enough salt in the mix. When this is done, take all of the set aside, clean cabbage leaves saved earlier, and lay flat on top of the sauerkraut, at least 3-4 layers thick of leaves. Place a round board that just fits into the crock on top of the leaves, and weigh down with a gallon jug full of water. Place crock in a cool place, preferably a basement. I usually put a tall kitchen trash bag over the top to make sure any stray bugs, etc. do not get into the crock. Fermentation time is normally 2½ to 3 weeks.

A check periodically will show that the bubbling action has come to a halt. The crock can then be opened up, the cover leaves discarded, and any amount of fermentation scum ladled off and thrown out. A good quality kraut should have a light amber color to it. Kraut is notorious for swelling during the water bath process, so we usually preheat the kraut before packing into jars. It is OK to add clean water to the kraut being heated. This will also tend to remove any excess saltiness. Only fill the jars about 80 percent full and add any brine necessary to cover the kraut in the jar. Processing time is in a boiling water bath for a period of 25-30 minutes. Then sit back and enjoy!

5 Easy Steps to Transition to a Whole Food Plant-Based Way of Eating

By Kellie Doiron

Embracing a Whole Food Plant-Based (WFPB) diet can be a life-changing decision, benefiting your health and the environment. This way of eating focuses on consuming whole, unprocessed plant foods while excluding animal products and refined items. If you're considering making the shift to a WFPB lifestyle, follow these five easy steps to help you get started on your journey to a healthier, more sustainable way of eating.

Step 1: Educate Yourself

Before embarking on your WFPB journey, it's essential to educate yourself about the principles of this lifestyle. Read books, watch documentaries, and explore online resources that provide insights into the health benefits, ethical considerations, and environmental impacts of a WFPB diet. Understanding the "why" behind your decision will help you stay motivated.

Choose one or two Plant-Based Dr's to follow. My favorites are:
- Dr McDougall https://www.drmcdougall.com/
- Dr Gregor https://nutritionfacts.org/

Step 2: Gradual Transition

Transitioning to a WFPB diet doesn't have to be an abrupt change. Start by incorporating more plant-based foods into your meals gradually. Begin with one or two meatless days a week and increase the number of plant-based meals over time. This gradual approach makes it easier to adapt and allows your taste buds to adjust to new flavors. Choose a date to transition completely when your ready.

Step 3: Explore Plant-Based Foods

The world of plant-based foods is incredibly diverse and delicious. Explore a variety of fruits, vegetables, legumes, whole grains, nuts, and seeds. Try new recipes and experiment with different cooking methods to keep your meals exciting. Foods like quinoa, lentils, tofu, and colorful veggies will become your new best friends.

Step 4: Meal Planning and Preparation

To ensure a successful transition, plan your meals ahead of time. Create a weekly meal plan, make a shopping list, and batch cook whenever possible. Having nutritious and tasty plant-based meals readily available will reduce the temptation to revert to old eating habits. Consider investing in cookbooks dedicated to plant-based cuisine for inspiration. Follow Plant-Based bloggers on Instagram, Tic Toc, or YouTube

Step 5: Stay Informed and Connected

Staying informed and connected with the WFPB community is vital for long-term success. Join online forums or social media groups where you can share experiences, ask questions, and discover new recipes. Attend local vegan or plant-based events to meet like-minded individuals who can offer support and guidance.

Conclusion:

Transitioning to a Whole Food Plant-Based way of eating is a rewarding journey that can improve your health, reduce your environmental impact, and promote animal welfare. By educating yourself, taking gradual steps, exploring new foods, planning your meals, and staying connected to the community, you can make this transition a smooth and sustainable process. Remember that every small change you make in your diet can have a significant impact, both on your well-being and the planet. So, take these five easy steps and start enjoying the benefits of a WFPB lifestyle today!

Kellie Doiron is a Whole Food Plant Based (WFPB) Nana of 2 teenage grandkids in Canada. She has been eating WFPB since September 2019 and could not imagine eating any other way. In the process, she has discovered a few recipes that she makes over and over, including her favorite hummus recipe (https://hellonutritarian.com/extra-creamy-no-oil-hummus) that can be found in her fridge most days. There are so many different flavours that can be incorporated with the chickpeas (Even chocolate!). You can find Kellie at kelliedoiron.com where she is an Associate for The Super Patch Company.

FOOD FOR THE APOCALYPSE

By Steven & Kathryn White

Part of being prepared for any emergency includes food storage. Whatever hits the fan - natural disaster, war, or a bad case of the munchies - you will have one less thing to worry about if you know you have plenty of food to get you through a crisis.

How much and which food you store will depend on the number of people in your household, your preferences, special health conditions, and space for storage.

Preparing a Three-Day Emergency Supply

A three day emergency kit is perfect for most disasters. Gathering the essential items that you might need and putting them in one location will help you get through the worst. This short-term kit should of course include food AND water (don't forget water!), personal hygiene items, flashlights, blankets and other essentials recommended for emergencies. The food you pack needs to be non-perishable; no refrigeration, minimal or no preparation or cooking, and little or no water to make (your water is precious in an emergency). For ease in managing your supply, select food items that are compact and lightweight. You may also want to pack away a small camp stove for boiling and heating.

Include a selection of the following foods in your short-term Disaster Supplies Kit:
- Ready-to-eat canned meats, fruits and vegetables
- Canned juices, milk, soup (if powdered, store extra water)
- Staples like sugar, salt, pepper
- High energy foods - peanut butter, jelly, crackers, granola bars, trail mix, etc.
- Foods for infants, elderly, or people on special diets
- Comfort/stress foods - cookies, hard candy, sweetened cereals, instant coffee, tea bags
- Vitamins

Make sure you have a can opener, scissors or knife for opening your food with, as well as disposable plates, cups and utensils. Pack all these items in plastic bags to keep them dry and as airtight as possible. Keep a list of dates when food items need to be inspected and rotated (used and then replaced with newly purchased items).

Foods in a refrigerator and freezer can be used at the beginning of the emergency, though frozen food in a well-insulated, well-filled, closed freezer will last for 2 to 3 days.

Preparing a Two-Week Emergency Supply

Even though it is unlikely that most emergencies will cut off your food supply for two weeks, some people have experienced this in recent years. If you are in an area where it is known that power can be off for extended periods, a two-week supply may seem very reasonable. The same general suggestions found above for a three-day supply will also work for a two-week supply.

One way to develop a two-week emergency supply is to increase the amount of basic foods you normally keep on your shelves. If you eat out regularly, take that into consideration. Your plan should not be based on foods that require being refrigerated or frozen. Many people already have a two-week supply of most staples on hand if they stop and think about it and make a written plan. Keep the supply fresh by rotating non-perishable items once or twice a year.

Suggestions to help you plan a two-week supply of food on hand:
1. Make a list of all family members by name, indicating any special needs (diabetic, allergies, etc.)
2. List all staple foods on your shelves now. Indicate amount available, date purchased, date opened and use by/replace date if known. Post this list near storage cabinets or closets and update when changes occur.
3. Repeat the step above for foods in your freezer.
4. Make a list of meals to be served, labeled "Day 1" through "Day 14". Indicate where food can be found for each day, if not stored all in one location.
5. Add notes to each day's list that indicate how much water and what equipment and utensils will be needed for preparation. This process will force you to think through what you will need to purchase and store.
6. Be sure to write up meal preparation steps or mix ratios on index cards and keep them closed in air- and water-tight plastic bags for use during the emergency.

Food Suggestions

Military and camping supply stores are good sources for some compact, well-preserved foods that are good choices for emergency kits. Dehydrated or freeze-dried foods are lightweight and take up little room, but you will need to plan extra water supplies for rehydrating them. If some foods in your kit require cooking, be sure to also include some that are ready to eat right away. Fires or stoves for cooking may be available during some emergencies, but you can't always count on that. In an emergency, what can go wrong, will go wrong, and ways to cook food are no exception!

Keep in mind that short-term emergency supplies need to emphasize survival, energy and hydration (water), but planning ahead means that you can also plan nutritionally balanced meals.

Food Ideas That Keep on the Shelf Ready to Eat:
- MRE's (meals-ready-to-eat)
- Shelf-stable "boxes" of juices and milk
- Crackers and melba toast (don't pick combination packs that require refrigeration)
- Dry, ready-to-eat cereals and granola
- Dried fruits, nuts and trail mixes
- Jerky
- Granola bars
- Hard candy

Dry Goods (plan water supplies to prepare):
- Dry beans and peas
- White rice
- Flour
- Pasta
- Rolled oats
- Instant potatoes
- Dry milk powder
- Powdered drink mixes
- Instant pudding
- Bouillon cubes or powder

Cooking supplies and condiments
- Sugar
- Salt
- Pepper
- Baking powder
- Baking soda
- Yeast
- Cocoa powder
- Olive oil
- Honey
- Maple syrup
- Molasses
- Vinegar - We use it for cleaning and disinfecting as well as cooking and canning.
- Vanilla extract - The real stuff (pure vanilla extract in alcohol) is expensive, but will keep for years.

Canned goods (some of these have oil or water in them that can be used for cooking)

- Canned tuna and/or salmon
- Canned meats
- Canned vegetables
- Canned fruit
- Canned soups
- Jams & Jellies
- Tomato sauce - Glass jars have a longer shelf life than cans for acidic products like tomatoes. And the jars are reusable.
- Canned beans
- Nut butter

Preparing a 3 Month Emergency Supply

If you want to stock a lot of calories for a long period for relatively little money, take the list of food above, measure how much each person will need for three months, vacuum pack it, and store it into 5 gallon buckets that can be sealed. Then, mark those buckets as for use only in a Mad Max or Book of Enoch style apocalypse where society is having to rebuild itself completely. Make sure you rotate the food buckets every 2-5 years!

Just the Facts - Politics from the horse's mouth…

I was treated to a two-week trip to Israel 32 years ago. It wasn't a commercial "guided tour," but a friend who had been there many times for extended stays paid my way over there. We shopped at the "farmers' market" in Jerusalem, stayed a week in the city, then rented a car and started in the Negev and crisscrossed our way north to the Lebanese border, Golan Heights, etc., staying in youth hostels and basically doing it "on the cheap."

While there I attended a sabbath celebration and spoke with a rabbi, who honestly told me that EVERY NEWS STORY coming out of Israel has FOUR VERSIONS:

1. The version they tell the world at large (middle east, etc.)
2. The version they tell the Israeli people.
3. The version they tell the Americans.
4. The truth, which nobody gets told.

If you are naïve enough to think the government there isn't lying every time they open their mouths (hmmm....sounds like our government, doesn't it?), then you are being played, just like the people you make fun of.

Michael Sawyer

Washington, August 29, 2023 - Following the U.S. Environmental Protection Agency (EPA) and the U.S. Department of the Army's announced final rule amending the on Waters of the United States (WOTUS), Glenn "GT" Thompson, Chairman of the House Committee on Agriculture, issued the following statement:

"America's farmers, ranchers, and landowners have come under regulatory assault from the Biden Administration. The Supreme Court rightfully curtailed EPA's regulatory overreach, however, EPA's bureaucratic sleight-of-hand circumventing the rulemaking process leaves the door open to agency abuse and regulatory and legal uncertainty for American agriculture. Our nation's farmers, ranchers, and landowners are excellent stewards of their land and understand that clean water and private property rights can coexist without burdensome overreach from the federal government."

Washington, October 25, 2023 - Senator James Lankford (R-OK) continues to raise concerns of "military-age, single adult men" coming across the border illegally. *"There are 70,000 people that are called 'special interests aliens' that the Administration has basically ignored and waived in,"* Lankford said. *"They've done fingerprinting on them and just allowed them in."*

According to Lankford, men from Iraq, Syria, Iran, and all over the Middle East are identified and released. *"They're from areas where terrorism is known to exist,"* Lankford said. *"But, they can literally travel anywhere they want. They're not being checked. They're not being monitored."*

Foreign Land Ownership Concerns Spark Changes on Congressional Committee

By Dawnita Fogleman

During a House Agriculture Committee hearing earlier this year, USDA Secretary Vilsack discussed the implications of a permanent placement of the USDA Secretary on CFIUS, saying, "Being a permanent member would allow us to educate the other members of CFIUS what to look for and what to be sensitive to when it comes to agriculture and agricultural production."

The Secretary of the Treasury chairs CFIUS. Members of the Committee include the heads of the Departments of Justice, Homeland Security, Commerce, Defense, State, Energy, the Office of the U.S. Trade Representative, and the Office of Science & Technology Policy.

"Protecting America's agriculture security is a critical part of our national security. With an increasing amount of foreign investment in U.S. agriculture, including the Secretary of Agriculture as a member of CFIUS is long overdue," said Congressman Frank Lucas. "I know firsthand just how important the security of our agricultural industry is, which is why I applaud my colleagues on the House Financial Services Committee in taking a critical step to make my longstanding priority law."

The Director of National Intelligence and the Secretary of Labor serve as non-voting members. While the Secretary of Treasury can designate the Secretary of Agriculture as a lead agency for transactions on a case-by-case basis, the Ag Secretary is not a permanent Member.

Agriculture is too important to go neglected. Mr. Lucas has long understood this, and I applaud his hard work in championing the Agriculture Secretary's contributions to CFIUS," stated Congressman Blaine Luetkemeyer, Chairman, House Financial Services Subcommittee on National Security, Illicit Finance, and International Financial Institutions.

The Agricultural Risk Review Act would add the Secretary of Agriculture as a permanent member to the Committee on Foreign Investment in the United States (CFIUS) for any transactions related to the purchase of agriculture land, agricultural biotechnology, or any other transaction related to the U.S. agriculture industry, as determined by the USDA Secretary.

"There is a small but growing share of agricultural land that is owned by foreign interests. Persons involved in those purchases as well as transactions affecting the food supply chain could very well pose a threat to the U.S." explained Congresswoman Maxine Waters, Ranking Member, House Financial Services Committee. "It is vital that CFIUS is equipped with relevant expertise to review transactions involving agricultural land or food-related businesses. H.R. 3378 is a commonsense bill that would formally add the Secretary of Agriculture to CFIUS, where food and agriculture are concerned."

How To Stay Connected During a Crisis

By Damian Allen

Disaster can happen at any place, at any time, and in times of crisis, communication becomes critical. Whether it's a natural disaster, a power outage, or a terrorist attack, being able to stay connected with loved ones, first responders, and emergency services is essential for survival. Unfortunately, during a crisis, traditional communication channels like phone lines and internet connections may become unavailable. That's why it's crucial to have a plan in place for emergency communications.

This is the digital prepper, and today, I'll be discussing the best ways to stay connected with friends, family, and other loved ones during a crisis.

Having a plan, and the technology involved

The first thing to consider when it comes to emergency communication is having a plan. You should have a predetermined meeting spot for your family in case you're separated during an emergency. You should also have a designated point of contact who can act as a liaison between your family and emergency services. This person should be someone who lives outside of the immediate area and who can be reached by phone or email.

Next, you should consider investing in a two-way radio. Two-way radios, also known as walkie-talkies, are a reliable form of communication during a crisis. They don't rely on cell phone towers or the internet, so they're more likely to work during a disaster. Two-way radios come in a range of sizes and prices, so you can find one that fits your needs and budget. Some models have a range of up to 35 miles, while others are more compact and can fit in your pocket.

Another option for emergency communication is satellite phones. Satellite phones are similar to cell phones, but they connect to satellites instead of cell phone towers. They work in remote areas where there's no cell phone coverage, and they're also a reliable option during a disaster. However, satellite phones are expensive, and the cost of the calls can be high.

If you don't want to invest in a two-way radio or a satellite phone, you can still use your cell phone for emergency communication. However, it's important to remember that cell phone networks can become overwhelmed during a disaster, and cell towers can be damaged or destroyed. That's why it's a good idea to have a backup power source for your cell phone, like a portable charger or a solar charger. You should also limit non-emergency phone calls and texts to conserve battery life and bandwidth.

Social Media can be useful, but listen to the news:

Another way to stay connected during a crisis is through social media. While traditional social media platforms like Facebook and Twitter may not be reliable during a disaster, there are other options available. For example, Nextdoor is a social media platform that's designed for neighborhoods. It allows you to connect with your neighbors and share information about local events, including emergencies. Other options include Zello, which is a walkie-talkie app that works over Wi-Fi and cellular data, and FireChat, which is a messaging app that works without an internet connection.

In addition to having a plan and the right equipment, it's also important to stay informed during a crisis. This means monitoring the news, listening to the radio, and following updates from local authorities. You should also consider signing up for emergency alerts from your local government or emergency services. These alerts can be sent to your cell phone, email, or two-way radio, and they provide information about evacuation orders, shelter locations, and other critical updates.

Practice, Practice, Practice!

Finally, it's important to practice your emergency communication plan before a crisis occurs. This means testing your two-way radios, charging your cell phone and backup batteries, and making sure you have the right contact information for your designated point of contact. You should also practice your plan with your family, so everyone knows what to do in case of an emergency.

In terms of Digital Preparedness, emergency communication is essential for staying safe and connected during a crisis. By having a plan in place and the right equipment, you can increase your chances of staying in touch with loved ones and emergency services. It's also important to remember that emergency preparedness involves more than just communication. Creating an emergency kit, knowing how to evacuate your home or workplace, and learning basic first aid are all important steps in staying safe during an emergency. By taking these steps, you can help ensure that you're prepared for whatever emergency may come your way.

Damian Allen is The Digital Prepper on YouTube. With well over a decade of Information Technology experience, his main goals are to try to explain to the prepping and homesteading community on how to be more prepared on a digital/technological level, "Digital Preparedness." Whether that be: Talking about online (or offline) based scams and how to deal with the aftermath, ways you can fortify your home network security, applications you can utilize to organize and maintain your home prepper inventory, or even the process of removing yourself from the internet entirely, these are things to keep in mind that you can utilize to help yourself, or prevent issues from bad actors.

The Best Gifts for New Moms

By Christayla Vassar

New moms can find the first few months of postpartum hard. So why not get them a gift that will make things a little easier for them? The following gifts aren't things for the most part, but services that you can either hire out for or give of your own time and ability.

Doula Service

Whether the new mom in your life has just had the baby or is going to, a doula service is the perfect gift.

Birth doulas provide emotional support for mothers-to-be, usually over text or email. They also provide physical support during the labor. It's important to note that they aren't medical providers, though. They won't be delivering the baby. Instead, they help the parents with counter pressure, massage, and more.

Postpartum doulas help after the baby has been born. They offer services such as debriefing the birth and the emotional state after it, housekeeping, preparing meals, watching the baby so mama can have a bath or a nap, and more.

Housekeeping

After the baby comes, it's hard to keep up with the house. Especially when a new mom is told to sleep when the baby sleeps! So why not give the gift of housekeeping? A new mom doesn't feel like taking care of laundry, cooking, or cleaning. After all, she just gave birth and is healing from a wound in her uterus, not counting any other wounds she may have.

When you go and help the new mom with laundry, dishes, and clutter or you hire a professional, you're helping in so many ways. You're allowing her time to rest, heal, and bond with her new baby. You're helping her not be overwhelmed by the mess.
If you want to offer housekeeping as a gift to a new mom, be sure to ask if it's something she would like.

Meal Train/Meal Delivery

Keeping 3+ meals on the table a day is hard when you have a new baby. Why not set up a meal train for an expecting or new mom?

There are websites you can use like Meal Train (mealtrain.com), Take Them A Meal (takethemameal.com), or Give In Kind (giveinkind.com). These websites help you organize a meal train for a new mom.

You can also help by bringing a new mom freezer meals. Whichever options you choose, check with her first to make sure there aren't any food allergies or tolerances.

Great nutrient-dense foods you can make for a postpartum momma:

- Congee
- Hot soups
- Cooked vegetables
- Protein and iron-rich casseroles
- Sweet potatoes and toppings
- Hot desserts (example: hot apple crisp)

Similarly, you can also give gift cards for meal delivery. Uber Eats, DoorDash, and GrubHub are three well-known meal delivery services you can use.

Errands

What's worse than realizing you're on your last roll of toilet paper? Realizing that you're on your last roll with no way to leave the house or a sleeping baby. You can offer to run errands when needed or even offer an Instacart or Walmart gift card. It takes a load off a new mom to have things delivered versus having to face the chaos of going to the store.

Teas

Teas help people relax and unwind. They are also a great way to replenish minerals and vitamins. A new mom can have many teas, but some are not safe postpartum.

Safe teas:
- Red raspberry leaf tea
- Chamomile
- Nettles
- Dandelion
- Fennel
- Lemon Balm
- Rooibos
- Green tea

Herbal Bath Teas

Herbal bath teas are a new trend that's so much fun! What is a better way to relax than a hot bath? You can find healing herbal bath teas as well. These herbal baths contain Epsom salts, sea salt, and herbs. You can find them on Etsy or use a recipe and make your own!

Snacks and drinks

Moms never experience hunger and thirst like postpartum hunger and thirst. A new mom will want to eat everything. Insatiable thirst will plague her every time she goes to nurse her baby. So why not get her a box of snacks she likes? Even better if she can eat them with one hand!

Ideas:

- Trail mix
- Nuts
- Popcorn
- Pretzels
- Tuna or chicken packets
- Olives
- Cottage cheese
- Snacking cheese
- Fresh-cut veggies or fruit
- Yogurt
- Protein bars
- Nut butter oat balls

Water, coconut water, and tea can help a mom stay hydrated. You can also help with add-ins like fruit or Liquid IV.

Diaper Stash or Fund

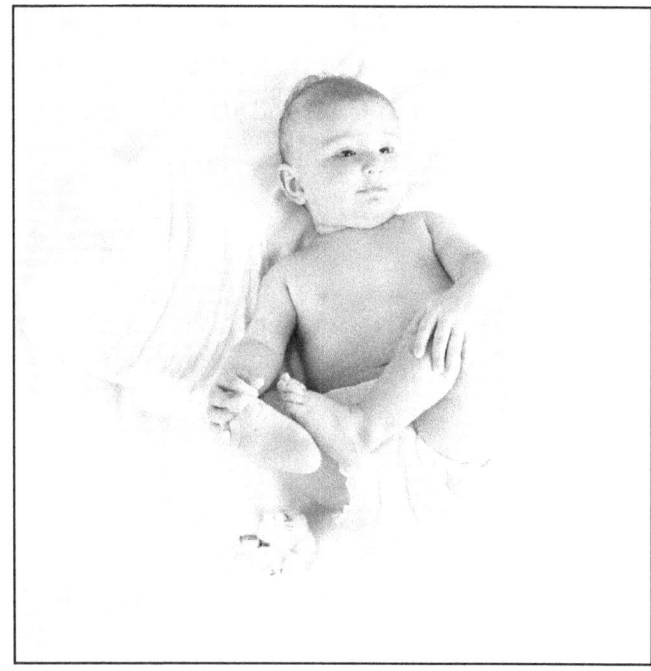

Diapers are expensive and babies grow fast. You can help by providing diapers or a cash/gift card for diapers! Ask the new mom what size would work best if you are bringing diapers, or bring diapers that are size 1 or above. Babies will be out of newborn diapers in no time! You could also help with wipes.

Another idea is cloth diapers on diaper-safe detergent. Cloth diapers can help save money in the long run by being used by many children. It's no longer a worry about safety pins either, as there are many other fasteners now!

Baby Carrier

There are many types of carriers. Some work best for newborns or young babies. Others work well for older babies and toddlers. Best of all, some can work for young babies to older toddlers!

The types of baby carriers:

- Ring sling - great for newborns to toddlers 25+ lbs
- Stretchy wrap - easy to wrap quickly; for newborns to 12 lbs.
- Woven wrap - great for wearing babies and toddlers up to 45 lbs! You can also wear more than one and wear in many different ways.
- Mei tai - great for newborns to 35 lbs.
- Soft structured carriers - these carriers work well for babies who have good head and neck control up to toddlers of 45 lbs. They can normally be worn on your front or back.
- Backpacks - great for active parents (hiking and mountain climbing) of older babies with good head and neck control.

A baby carrier can help a mom get stuff done after she heals and makes a great gift!

Of course, there are many other gifts you can give a newly postpartum mom. Be creative! And thank you for being her village!

Preparing for a VBAC

By Emmie Manor

Childbirth is an amazing event that YHWH has given to women. In many ways it can be a hot topic with all of the talk of birth control, over-population, and many other lies that are circulating in the culture today. But YHWH makes it clear in His Word that giving birth is what women were created to do and it is a blessing.

"Behold, children are a gift from YHWH, The fruit of the womb a reward." Psalm 127:3

While this amazing time is natural and created by YHWH, preparation must be done. Much of the preparation for childbirth is the same, whether it's your first or tenth baby and whether you're planning for a vaginal birth, C-section or Vaginal Birth After Cesarean (VBAC).

For example, no matter what you need to be deep in prayer and in the Bible seeking YHWH and His best for your baby and you. It is imperative that you are connected with your Creator, growing in relationship with Him, telling Him your desires and hearing His voice. Give Him your fears, your doubts, your questions and let Him speak Truth over you.

As you study the Word of YHWH memorize verses and passages, especially a Psalm or two. If you're struggling to memorize, or even if you're not, write down your favorite verses and passages to be read over you during your birth.

You also need to seek wisdom from those who have gone before you. Medical professionals have their experience and knowledge that is very valuable. But just as valuable is talking to "older women" (Titus 2) who have given birth and asking them their experience as well as how they made the decisions they did for their birth or learning about emergent situations and how they walked through it. This type of preparation is priceless.

Talking with your husband is also needed. Tell him what you're thinking, feeling, wanting and let him tell you his thoughts. Seek YHWH together in prayer and Bible reading and listen to His voice together. Childbirth can be an amazingly unifying time if you are seeking YHWH and growing in Him together as your baby grows in you.

Eating nourishing food that is nutrient filled, taking needed supplements and drinking hydrating fluids is also extremely important. This helps your body prepare for the labor as well as gives your baby an amazing start. Proteins, healthy fats and carbs along with vitamins and minerals build a strong body for the baby and you.

My favorite prenatal vitamin is from the store Sprouts if you have one in your area:
- **Sprouts Organic Prenatal Whole Food Vitamin**
- **Whole Food Vitamin:**
  https://www.vitacost.com/garden-of-life-vitamin-code-raw-prenatal-180-vegetarian-capsules

Other options that I've used though not specifically prenatal, but still a whole food vitamin:
- **Amare Vita GBX** https://www.amare.com/60675/en-US/VitaGBX
- **Plexus X Factor or X Factor Plus** https://plexusworldwide.com/emmiemanor/home

Along with healthy food, being active and getting regular exercise is also needed to prepare for birth. Exercising your abs, legs and pelvic floor are important to be ready for whatever your labor might bring. Taking mild walks, swims and ab exercises that encourage strength and keep your abdominal muscles together are all beneficial for growing strength.

All of these things apply no matter what type of birth you have. However there are other considerations needed for you if you are planning a VBAC.

First, you need to add extra vitamins and supplements to your regimen. Because of the trauma the skin, muscles and uterus have been through before, you need to consume extra protein to build the fibers in every layer. Some of the best sources are beef collagen, homemade bone broth, and meat sources, locally bought is best but organic, grass fed is still good. If both are out of your price range, get what you can.

- (My favorite beef collagen and the one I've found for the best price comes from Trim Healthy Mama. You can find it here:

  https://store.trimhealthymama.com/product/integral-collagen/)

You also want to increase your folic acid or folate more than what's recommended for pregnancy. This B-vitamin helps with healing scar tissue and strengthens the areas that were cut into before. Getting a whole food supplement or vitamin is the best while also increasing foods with folate in it. Foods high in folate include: dark green leafy vegetables (spinach, turnip greens, Brussels sprouts, broccoli, asparagus, etc), liver, lentils, beans, avocado, beets and citrus fruits. There are other foods also, but this will get you started.

- (Whole Food B Complex:

  https://www.vitacost.com/garden-of-life-vitamin-code-raw-b-complex)

Vitamin E whole food supplement and foods are also beneficial for strengthening the skin and muscles. Some good sources for vitamin E are almonds, sunflower seeds, avocado, peanut butter, red bell peppers and fish.

- (Whole Food E Vitamin:

  https://www.vitacost.com/natural-factors-whole-earth-sea-sunflower-vitamin-e)

It's also a good idea to use a healing oil on your skin where the previous scar is. There are many oils that would be beneficial including castor, olive and coconut among others.

- (This is a great source of castor oil in a roll on so it's easy to apply:

  https://www.amazon.com/Heritage-Nourishing-Treatment-Hydration-Application/dp/B0013ZFU2O/)

It would also be beneficial to do extra pelvic floor and abdominal exercises designed for pregnancy. Combined with added supplements, especially folate/folic acid, this will help build strength and reduce scar tissue from the previous c-section and prepare your body well for a vaginal birth.

Spending extra time in prayer and Bible study would also be extremely beneficial. This is especially true if your c-section was unplanned or traumatic in any way. You may have had in your mind exactly how your birth was going to happen and then things happened outside of your control that required a c-section. Many feelings can come with that. If you are dealing with negative feelings and emotions or any fear from that situation, then please take as much time as you can to pray over this, taking your cares to YHWH and talking with other women that have walked the same road. If you don't know someone, start asking around! There are many who have gone before you that can help you walk this road.

Along with extra time in prayer and Bible study, extra conversations with your husband may be needed. Depending on the circumstances of the previous c-section, he may have concerns that need to be addressed also. Pray together and seek YHWH's wisdom in all of these things.

Nothing is a guaranteed, however, and so the biggest part of preparation is knowing that it is possible something could happen that would require another C-section. Do not let this be your primary focus. Focus on YHWH and sharing your desire with Him. Focus on getting your body in the best shape you can, being wise and healthy with food choices and exercise so that you are ready for a vaginal birth. But also, know, acknowledge, that a repeat c-section could happen and leave it in the hands of your Savior.

I'm praying for you, Sister! Praying that YHWH will give you the desires of your heart and that it will be a beautiful and amazing birth.

Jim and Emmie Manor are followers of Christ, Bible believers and passionate about family discipleship. They have been married for 17 years and are the parents of 11 children ages 12 and under.

Both Jim and Emmie are Registered Nurses and Jim has a Masters of Divinity from Southern Seminary. Between seminary studies and our growing family and life experiences, we have learned much about discipling children at different ages and stages and how to (and not to) reach children with the Gospel.

Currently Jim is working as an RN and Emmie is a homemaker who homeschools their children and writes.

Gift Ideas for Preppers

By Steven and Kathryn White

If you don't know who Preppers are, they are people who prioritize preparedness for various emergencies and unexpected situations. When choosing gifts for preppers, it's essential to consider items that can aid in their self-reliance, survival, and overall preparedness. Here are ten gift ideas for preppers:

- **Multi-Tool or Swiss Army Knife:** A high-quality multi-tool or Swiss Army knife with various functions, such as a blade, can opener, screwdriver, and more, is a versatile and essential tool for any prepper.
- **Emergency Food Supplies:** Consider gifting freeze-dried meals, MREs (Meals Ready to Eat), or long-lasting food items like canned goods, to help build their emergency food stash.
- **Water Filtration System:** A portable water filter or purification system, like a LifeStraw or a gravity water filter, can provide access to clean and safe drinking water during emergencies.
- **First Aid Kit:** A comprehensive first aid kit with essential medical supplies and equipment is a vital item for any prepper. Look for kits that include bandages, antiseptics, and basic medical tools.
- **Fire Starter Kit:** Fire-starting tools like a Ferro rod, waterproof matches, or a reliable fire starter can be invaluable for cooking, warmth, and signaling for help in survival situations.
- **Flashlights and Headlamps:** High-quality, durable LED flashlights and headlamps are essential for preppers to navigate in the dark, signal for help, or simply perform tasks during power outages.
- **Camping Gear:** Camping equipment like a lightweight tent, sleeping bag, portable stove, and cookware can serve double duty for both recreational camping and emergency situations.
- **Solar Charger or Power Bank:** A portable solar charger or power bank can keep their electronic devices charged, ensuring they have access to communication, maps, and information during emergencies.
- **Survival Books and Guides:** Preppers often value knowledge and skills. Consider gifting survival guides, books on foraging, wilderness first aid, or any other relevant topic to enhance their preparedness.
- **Security and Self-Defense Tools:** Non-lethal self-defense tools like pepper spray, personal alarms, or a tactical flashlight with a stun function can provide peace of mind and security during uncertain times.

When selecting gifts for preppers, be sure to consider their specific needs and preferences. Whether they focus on urban survival, wilderness survival, or a mix of both, these gift ideas can help them enhance their preparedness and resilience in the face of unexpected events.

4-H History Project Ideas

By Myrna Buckles

Expressive & Communication Arts Ideas:

- If your family has been involved in organizing an event in your community, gather photos and stories of past events as well as information about how it got started and why. Document the history of the event on a posterboard and make an oral presentation.
- The child/teen can record stories told by their grandparents and transcribe them into a simple bound book. Be sure to include photos. This could also be done as a poster and oral presentation.
- Find a historic building in your community, learn about its history and construct a miniature version of it. For instance, this is the historic home I grew up in. A student built this model of it with my dad's help. This is a tourist's photo when my dad gave them a tour of his museum.

Plant Sciences Ideas:

- Ask a parent or grandparent to teach you how to grow a certain type of vegetable or design a flower garden. Enter your produce in the local fair.
- Plant a garden with heirloom seeds and learn about their history. I know of at least one museum that sells heirloom seeds and teaches about their history. (The Museum of the Fur Trade at Chadron, Nebraska found online at https://www.furtrade.org/) Enter your produce in the local fair and also prepare a poster presentation.

Animal Science Ideas:

- Learn the history of the breed(s) of animals you plan to show at the local fair. Prepare a poster presentation about one of them.

Family and Consumer Science:

- Ask a grandparent or parent to teach you how to make a baked good recipe that has been in your family for generations. Practice making the recipe so you can enter your baked goods in the fair.
- Ask a grandparent or parent to teach you to sew, knit, crochet, cross stitch, or embroidery and make a project to enter in your local fair.
- Gather family heirloom recipes and make your own family cookbook.

Natural Resources:

- Learn about fishing methods of the past and prepare a poster presentation.
- Learn the history of the gun you most like to hunt with and write a report. When was it first made, who designed it, who is the manufacturer and how long has it been being made?
- Learn about fossils in your area or nearby. Prepare a presentation.

These are just a few options for possible 4-H projects. Start now so you are well prepared at the time of your local fair and have fun learning.

From Struggle to Strength: High School and Learning Disabilities

By PJ Pitonyak, M. Ed.

High school. Do those two words bring back joyful memories or painful ones? For me who is coming up on my 45th year high school reunion, I can look back and smile because school was easy for me. I can, however, think of many people who may not attend the reunion. One reason may be that they did not enjoy school because learning in that environment was difficult for them. High school can be hard enough without having learning challenges, but having a diagnosed learning disability can make those years much more difficult. I do, however, have some hope to offer.

Let's unpack the previous paragraph. Notice that I used the term "diagnosed learning disability." That is one key to opening doors for a struggling learner. If you are reading this, you may be thinking about your own child. Is he or she in high school? Are they struggling with academics? Have they been diagnosed by professionals? If yes, were they diagnosed when at the elementary level? That would have been ideal. However, many students are not diagnosed until middle school and some not until high school. No matter when, help can be made available. The point is to get a formal diagnosis. Without a formal diagnosis, learning and passing classes can be more of a challenge than necessary. Teachers, the special education department in your school system, your doctor, and a psychiatrist can help with this. Even if a child is homeschooled, a parent can and should reach out to the public school system to find help.

When one hears the term "learning disability," the first word that comes to mind often is dyslexia, but there is so much more to understand about this disorder. To begin with, a person with a learning disability is not necessarily disabled. He or she is unique and learns differently. Secondly, there are many forms of learning challenges in addition to dyslexia. Dyslexia is a language processing disorder affecting reading, writing, and comprehension. Dysgraphia pertains to challenges in writing, while dyscalculia hinders one's ability to work effectively with numbers. A person may have one but often has more than one disability. Another way to group them is by spoken language including listening and speaking, written language including reading, writing, and spelling, and arithmetic including calculation and concepts. They all fall under one umbrella, and a person can have one, many, or all disabilities. In addition, one may have Attention-deficit/hyperactivity disorder (ADHD) as well, but that is for another article. One may also show signs of depression. In high school aged males, this can be expressed through anger.

Understanding a learning disability is just the beginning because there are often other factors to consider.

Now that we have a better understanding of what a learning disability is, it is important to understand what it is not. It is not an excuse for misbehaving in school or skipping school or quitting school. There are teachers and counselors at school who can help. There are formal Individual Education Plans (IEPs) that provide accommodations and modifications such as extra time, use of a calculator, use of a word bank, use of a word processor or computer, work being broken down into smaller chunks, checks for understanding, alternative assignments, small group instruction, and more.

Know your child's rights to education. If homeschooling, use all accommodations and modifications as possible to ensure successful learning and schoolwork.

Let's just highlight that schools are totally different from when I was in high school from 1974 to 1978. Today we have technology that helps spell and write and read and calculate. There really is no longer a reason not to enjoy the classics or not to be able to write ideas when they can be spoken into a word processor and edited with an artificial intelligence (AI) device.

Math weaknesses may be harder to solve (Pun intended.) because one needs to understand the concepts and formulas to plug into the calculator the correct numbers in the correct order with the correct operation, but one can find lessons on YouTube for free to teach math concepts and practices.

The Internet is full of free and paid programs, many in the form of games, that can help with any academic need. Tutoring can be done in person as well as through programs like Zoom and Class-In. The sources of help are endless.

High school students look to the future, sometimes with anxiety, but I want to offer encouragement that their choices are limitless. With tools and accommodations, students can plan to attend college or trade school or go straight to work. It is important to be honest with one's skills and lack of skills when applying to programs. Look for programs that offer help. IEPs go to the age of 21.

There are other resources that are also offered past graduation, but there are qualifications and/or restrictions. Research them before graduation. One's chances of joining the military may be limited, but each person can investigate that for himself if interested.

Let's end with some thoughts from the BIBLE.

> *"So God created man in his own image, in the image of God created he him; male and female created he them." Genesis 1:27*

God does not make mistakes. A learning disability does not have to be crippling. He can give you help and strength.

> *"For the Lord is a God of Knowledge." 1 Samuel 2:3 (KJV)*

He has given us many modern-day tools with which to learn, even with a learning disability.

> *"And he said, The things which are impossible with men are possible with God." Luke 18:27*

Each person has an ability to learn in his or her way to the best of his or her ability. There are many ways to help each learner.

God has given us a world full of wonder, and with His Grace and our grace while learning, we can share in great experiences and knowledge, having peace of mind because we know we have done our best to learn and do and honor Him. There is a wealth of information at our fingertips, empowering us to embrace learning and overcome challenges with confidence.

Patricia Jean (PJ) Pitonyak is a retired special education teacher with over 41 years of experience. She now teaches English as a Second Language from her home in southern Maryland to students in over 8 countries. She is happily married to her husband of 27 years and has many children and grandchildren. She is the recipient of the 10th Annual Award for Individual Secondary Winner from the Special Education Citizens Advisory Committee (SECAC). For more information or a consultation, contact her at pj@wonderandgracelifecoaching.com and visit her website at https://wonderandgracelifecoaching.com.

To every thing there is a season,
and a time to every purpose under the heaven:
A time to be born, and a time to die;
a time to plant,
and a time to pluck up that which is planted;
A time to kill, and a time to heal;
a time to break down, and a time to build up;
A time to weep, and a time to laugh;
a time to mourn,
and a time to dance;
A time to cast away stones,
and a time to gather stones together;
a time to embrace,
and a time to refrain from embracing;
A time to get,
and a time to lose;
a time to keep,
and a time to cast away;
A time to rend,
and a time to sew;
a time to keep silence,
and a time to speak;
A time to love,
and a time to hate;
a time of war,
and a time of peace.

Ecclesiastes 3:1-8 (KJV)

Subscribe to this quarterly e-publication at PrairieDustTrail.com

Dawnita Fogleman is a fifth-generation Oklahoma Panhandle homesteader, homeschool mom of six, author, and award-winning journalist, connecting the past with the future to help you weather the dust storms of life with God's Divine help.

@PrairieDust on YouTube
Patreon = bit.ly/Prairie-Dust-Patreon
@dawnitafogleman on Instagram, Twitter & Gab
@prairie_dust on Instagram and Twitter and TikTok
PrairieDustTrail.com

www.ingramcontent.com/pod-product-compliance
Lightning Source LLC
Chambersburg PA
CBHW082133290526
45794CB00008B/3021